T0237163

**Tutorials, Schools, and Workshops
in the Mathematical Sciences**

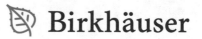

This series will serve as a resource for the publication of results and developments presented at summer or winter schools, workshops, tutorials, and seminars. Written in an informal and accessible style, they present important and emerging topics in scientific research for PhD students and researchers. Filling a gap between traditional lecture notes, proceedings, and standard textbooks, the titles included in TSWMS present material from the forefront of research.

More information about this series at http://www.springer.com/series/15641

David Henry • Konstantinos Kalimeris •
Emilian I. Părău • Jean-Marc Vanden-Broeck •
Erik Wahlén

Editors

Nonlinear Water Waves

An Interdisciplinary Interface

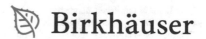 Birkhäuser

Editors
David Henry
School of Mathematical Sciences
University College Cork
Cork, Ireland

Konstantinos Kalimeris
Department of Applied Mathematics
and Theoretical Physics
University of Cambridge
Cambridge, UK

Emilian I. Părău
School of Mathematics
University of East Anglia
Norwich, UK

Jean-Marc Vanden-Broeck
Department of Mathematics
University College London
London, UK

Erik Wahlén
Centre for Mathematical Sciences
Lund University
Lund, Sweden

ISSN 2522-0969   ISSN 2522-0977   (electronic)
Tutorials, Schools, and Workshops in the Mathematical Sciences
ISBN 978-3-030-33538-0   ISBN 978-3-030-33536-6   (eBook)
https://doi.org/10.1007/978-3-030-33536-6

Mathematics Subject Classification (2010): 76B15, 76B25, 76B55, 76B03, 76U05, 35J57

© Springer Nature Switzerland AG 2019
This work is subject to copyright. All rights are reserved by the Publisher, whether the whole or part of the material is concerned, specifically the rights of translation, reprinting, reuse of illustrations, recitation, broadcasting, reproduction on microfilms or in any other physical way, and transmission or information storage and retrieval, electronic adaptation, computer software, or by similar or dissimilar methodology now known or hereafter developed.
The use of general descriptive names, registered names, trademarks, service marks, etc. in this publication does not imply, even in the absence of a specific statement, that such names are exempt from the relevant protective laws and regulations and therefore free for general use.
The publisher, the authors, and the editors are safe to assume that the advice and information in this book are believed to be true and accurate at the date of publication. Neither the publisher nor the authors or the editors give a warranty, express or implied, with respect to the material contained herein or for any errors or omissions that may have been made. The publisher remains neutral with regard to jurisdictional claims in published maps and institutional affiliations.

This book is published under the imprint Birkhäuser, www.birkhauser-science.com, by the registered company Springer Nature Switzerland AG.
The registered company address is: Gewerbestrasse 11, 6330 Cham, Switzerland

Preface

The motion of water is governed by a set of mathematical equations which are highly complicated and intractable, which is not surprising when one considers the highly diverse and intricate physical phenomena which may be exhibited by a given body of water. However, recent mathematical advances have enabled researchers to make vast progress in this field. Cutting-edge techniques and tools from mathematical analysis have generated strong rigorous results concerning the qualitative and quantitative physical properties of solutions of the governing equations. Furthermore, accurate numerical computations of fully nonlinear steady and unsteady water waves in two and three dimensions have contributed to the discovery of new types of waves. Model equations have been derived in the long-wave and modulational regime using Hamiltonian formulations and solved numerically.

Additionally, while research in nonlinear water waves has an inherent symbiotic relationship with the generation of powerful mathematical advances, it is also a subject which has vast potential for interdisciplinary collaborations. In many instances throughout science, a numerical or experimental exploration is the first approach to obtaining important information about the behaviour of solutions of differential equations. Conversely, behaviour which is predicted by mathematical theory may subsequently be observed and expanded upon in experimental or numerical work. The aim of this book is to reflect, and illustrate, the wide variety of approaches to the analysis of nonlinear water waves, featuring a range of exponents of recent research in the theory and applications of nonlinear water waves.

This book is motivated by a workshop on nonlinear water waves which was organised at the Erwin Schrödinger International Institute for Mathematics and Physics in Vienna, November 27–December 7, 2017, and features contributions from a number of participants. The aim of the workshop was to examine recent progress in the research area of nonlinear water waves with a view to stimulating future research and collaborations. International experts in the broad domain of fluid dynamics were present, with featured interdisciplinary expertise ranging from pure and applied mathematicians to physicists, oceanographers, experimentalists, and engineers.

The workshop was organised around five main themes: Nonlinear wave–current interactions; Geophysical water waves; Analysis and justification of asymptotic models for water waves; Numerical computations of water waves; Nonlinear surface waves in related physical problems. Furthermore, the material featured in the chapters of this volume range from new research results to review articles outlining the state-of-the-art of a particular field. Accordingly, it is hoped that this volume will be of interest to both experts, and early career researchers, alike. An overview of the volume is given as follows.

In Chap. 1, Nachbin introduces some applied mathematics research problems on surface water waves propagating in the presence of highly variable bottom topographies. One problem regards solution properties, in particular connected with wave reflection, while another asymptotic problem regards reduced models obtained by simplifying partial differential operators, and which are more amenable to scientific computing.

Kluczek and Rodríguez-Sanjurjo, in Chap. 2, implement a degree-theoretic approach in order to rigorously justify the global validity of the fluid motion described by a new, exact, and explicit solution of the nonlinear geophysical fluid dynamic governing equations. More precisely, the three-dimensional Lagrangian flow-map describing this exact and explicit solution is proven to be a global diffeomorphism from the labelling domain into the fluid domain, thereby establishing that the flow motion is dynamically possible.

In Chap. 3, Fokas and Kalimeris review the application of the unified transform, also known as the Fokas method, to the water waves problem. The Fokas method involves a non-local formulation, and its application in water waves in two spatial dimensions with moving boundaries is discussed.

Guyenne presents in Chap. 4 an overview of recent extensions of the high-order spectral method of Craig and Sulem. Cases of wave propagation in the presence of fragmented sea ice, variable bathymetry, and a vertically sheared current are investigated. The main components of this method include reduction of the full problem to a lower-dimensional system involving boundary variables alone, and a Taylor series representation of the Dirichlet–Neumann operator, resulting in a very efficient and accurate numerical solver by using the fast Fourier transform.

In Chap. 5 Dyachenko and Hur investigate the Stokes wave problem in a constant vorticity flow by using a numerical method based on a formulation via a conformal mapping as a modified Babenko equation. Touching waves of different types are found by varying the strength of the vorticity. A fold is found to develop in the wave speed versus amplitude plane for strong positive vorticity.

In Chap. 6, Compelli, Ivanov, and Lyons consider a model from physical oceanography, namely a two-layer fluid system separated by a pycnocline, which is modelled by an internal wave. The lower layer is bounded below by a flat bottom, and the upper layer is bounded above by a flat surface. The fluids are incompressible and inviscid and Coriolis forces as well as currents are taken into consideration. A Hamiltonian formulation is presented, and appropriate scaling leads to a KdV approximation. Additionally, considering the lower layer to be infinitely deep leads to a Benjamin–Ono approximation.

Akers and Seiders present in Chap. 7 numerical simulations of large amplitude overturned travelling waves using a dimension-breaking continuation as a numerical technique. They present dimension-breaking bifurcations from branches of planar waves in two weakly nonlinear model equations as well as in the vortex sheet formulation of the water wave problem. The challenges and potential of this method toward computing overturned travelling waves at the interface between three-dimensional fluids are reviewed and numerical simulations of dimension-breaking continuation are presented in each model.

In Chap. 8, Bauer, Cummings, and Schneider consider a model for the periodic water wave problem and its long wave amplitude equations. The validity of the KdV and of the long wave NLS approximation for the water wave problem over a periodic bottom is investigated.

Amann reviews in Chap. 9 some recent numerical methods used to compute steady periodic water waves in two dimensions. These different methods are based on a Dubreil–Jacotin transformation, on a non-local formulation, and on conformal mapping.

Chapter 10 comprises a review by Stuhlmeier, Vrecica, and Toledo of the theory of wave interaction in finite and infinite depth, with a focus on coastal engineering applications. Both of these strands of water-wave research begin with the deterministic governing equations for water waves, from which simplified equations can be derived to model situations of interest, such as the mild slope and modified mild slope equations, the Zakharov equation, or the nonlinear Schrödinger equation. These deterministic equations yield accompanying stochastic equations for averaged quantities of the sea-state, like the spectrum or bispectrum. The authors then discuss several of these in depth, touching on recent results about the stability of open ocean spectra to inhomogeneous disturbances, as well as new stochastic equations for the nearshore.

In Chap. 11, Părău and Vanden-Broeck review the solitary gravity-capillary and flexural-gravity waves in two and three dimensions found over the years by water waves researchers. The numerical methods used to compute these solitary waves are described in detail and similarities and differences between the solutions for the two physical problems are discussed.

Trichtchenko presents in Chap. 12 the details of a method for identifying stability regimes of small-amplitude, periodic travelling wave solutions of dispersive Hamiltonian partial differential equations to high-frequency perturbations using roots of a reduced-order polynomial.

Finally, the editors and authors would like to acknowledge the Erwin Schrödinger International Institute for Mathematics and Physics, Vienna, for their immense support and hospitality during the 2017 workshop on *Nonlinear Water Waves: An*

Interdisciplinary Interface. This workshop was an engaging forum for discussions and interactions between all scientific researchers in attendance and in the local community, and all participants greatly appreciated the facilities and support of the institute and its staff.

Cork, Ireland   David Henry
Cambridge, UK   Konstantinos Kalimeris
Norwich, UK   Emilian I. Părău
London, UK   Jean-Marc Vanden-Broeck
Lund, Sweden   Erik Wahlén

Contents

Modeling Surface Waves Over Highly Variable Topographies

André Nachbin

Abstract This article introduces some applied mathematics research problems on surface water waves propagating in the presence of highly variable bottom topographies. Asymptotic problems arise from variable coefficient partial differential equations regarding both its solutions, as well as the differential operators' reduced modeling. Two simple problems are first introduced, setting the main ideas.

Keywords Water waves · Variable topographies · Reduced modeling · Effective solutions

Mathematics Subject Classification (2000) Primary 76B15; Secondary 35Q

1 Introduction

Water waves is a subject of great current interest regarding both the continuous mathematical challenges it poses, as well as the many physical applications of interest. A topic which is particularly challenging, theoretically and numerically, is water wave interaction with the bottom topography. The goal of this article is to introduce the reader to this class of applied math problems in water waves. As will be shown, very quickly the problems become quite involved. Therefore we start with two simpler problems, in the respective subareas considered, all on the topic of partial differential equations applied to waves. One problem regards solution properties, in particular connected with wave reflection. The other problem regards reduced models, namely on simplifying partial differential operators in order to obtain models amenable to the asymptotic analysis of their solutions, as

---

This work was completed with the support of our TEX-pert.

A. Nachbin (⊠)
Institute for Pure and Applied Mathematics (IMPA), Rio de Janeiro, RJ, Brazil
e-mail: nachbin@impa.br

© Springer Nature Switzerland AG 2019
D. Henry et al. (eds.), *Nonlinear Water Waves*, Tutorials, Schools, and Workshops
in the Mathematical Sciences, https://doi.org/10.1007/978-3-030-33536-6_1

well as more amenable to scientific computing. Many interesting phenomena and theoretical results, here discussed, are inspired from acoustic waves. We start with a very simple linear hyperbolic (acoustic) model which sets this parallel.

This article is organized as follows. In Sect. 2 we present three different settings for the acoustic wave reflection problem. The analogy with linear shallow water waves is explored. In Sect. 3 we describe issues related to reduced modeling in the presence of a rapidly varying propagation medium. In Sect. 4 we present an overview of more advanced developments regarding recent research on these topics.

2   Long Wave Reflection

We start this section with a simple exercise for a linear hyperbolic wave reflecting at an interface. It is instructive to consider the linear acoustic equations

$$\rho(x)\, u_t + p_x = 0, \tag{2.1}$$

$$\frac{1}{\kappa(x)}\, p_t + u_x = 0.$$

It is an easy exercise to adapt the calculations here presented for the acoustic equations to the shallow water model presented in Eq. (2.4). Here we consider a one-dimensional heterogeneous acoustic medium along the x-axis. The material density is given by $\rho(x)$, where $\rho(x) = \rho_1$ when $x < 0$ and $\rho(x) = \rho_2$ when $x > 0$. The material compressibility is given through the variable coefficient $1/\kappa(x)$, which is also discontinuous at $x = 0$, due to two different constant values of the bulk modulus, $\kappa_1 \neq \kappa_2$. The velocity at a point x in the variable medium is denoted by $u(x, t)$, while $p(x, t)$ is the pressure. We will choose initial conditions so that a pulse-shaped disturbance in pressure and velocity will travel from left to the right, and interact with the medium's interface at $x = 0$.

2.1   Reflection at an Interface or Step

Using the method of characteristics it is straightforward to compute the reflection and transmission coefficients at the interface where the medium changes acoustic wave speed. One can write system (2.1) in matrix notation and find the matrix's eigenvalues which are the right and left wave speeds: $c(x) = \pm[\kappa(x)/\rho(x)]^{1/2}$. This system of equation is therefore bi-directional supporting waves in either the right or left going directions. Right and left going modes are defined by

$$A(x, t) \equiv \zeta^{-1/2}\, p(x, t) + \zeta^{1/2}\, u(x, t) \quad \text{and} \tag{2.2}$$

$$B(x, t) \equiv -\zeta^{-1/2}\, p(x, t) + \zeta^{1/2}\, u(x, t),$$

respectively, which when substituted in system (2.1) yield

$$A_t + c\, A_x = -r(x)\, B,$$ (2.3)
$$B_t - c\, B_x = r(x)\, A.$$

The reflectivity coefficient $r(x) \equiv \zeta'/(2\zeta)$, which depends on the impedance $\zeta(x) \equiv (\kappa\rho)^{1/2}$, couples the right and left propagating modes when the medium changes. If the medium properties are constant the two equations decouple into elementary unidirectional wave equations, in opposite directions. A and B are an elementary version of a Riemann Invariant along a characteristic. When the medium properties change they are no longer invariant and exchange energy through the reflectivity terms. Note that a matched-medium situation can occur even when the propagation speeds change: $c_1 \neq c_2$. This occurs when the impedances match, as for example with $\rho_1/\rho_2 = \kappa_2/\kappa_1$. This will be confirmed in a calculation presented below.

Consider the very simple linear shallow water model. Note that we do not have a matched-medium possibility in this case:

$$u_t + h(x)\eta_x = 0,$$ (2.4)
$$\eta_t + u_x = 0.$$

We have only one variable coefficient in this system. But the calculations that follow all carry out to the linear shallow water case.

Consider the single interface case depicted in Fig. 1, to the left. In order to calculate the reflection and transmission coefficients we impose continuity of the acoustic solution at the interface located at $x = 0$:

$$u(0, t) = \zeta^{-1/2}\left(\frac{A_1(0, t) + B_1(0, t)}{2}\right) = \zeta^{-1/2}\left(\frac{A_2(0, t) + B_2(0, t)}{2}\right),$$ (2.5)

$$p(0, t) = \zeta^{1/2}\left(\frac{A_1(0, t) - B_1(0, t)}{2}\right) = \zeta^{1/2}\left(\frac{A_2(0, t) - B_2(0, t)}{2}\right).$$ (2.6)

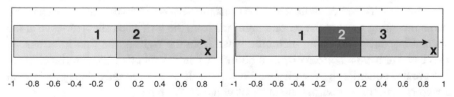

Fig. 1 Left: single interface between medium 1 and 2. Right: layer of medium 2 located in between medium 1 and 3

This leads to the propagator matrix, taking the solution across the interface from side 1 to side 2:

$$\begin{bmatrix} A_2 \\ B_2 \end{bmatrix} = \begin{bmatrix} p^+ & p^- \\ p^- & p^+ \end{bmatrix} \begin{bmatrix} A_1 \\ B_1 \end{bmatrix}, \tag{2.7}$$

where $p^+ \equiv [(\zeta_2/\zeta_1)^{1/2} + (\zeta_1/\zeta_2)^{1/2}]/2$ and $p^- \equiv [(\zeta_2/\zeta_1)^{1/2} - (\zeta_1/\zeta_2)^{1/2}]/2$. Let's admit a disturbance arriving at the interface from the left and nothing arriving from the right. Our boundary conditions are therefore $A_1(0^-, t) = f(t)$ and $B_2(0^+, t) = 0$. The propagator relations reads

$$\begin{bmatrix} A_2 \\ 0 \end{bmatrix} = P \begin{bmatrix} f(t) \\ B_1 \end{bmatrix}, \tag{2.8}$$

and we get the reflection and transmission coefficients

$$R = -\frac{p^-}{p^+} = \frac{\zeta_1 - \zeta_2}{\zeta_1 + \zeta_2}, \quad T = \frac{1}{p^+} = \frac{2\sqrt{\zeta_1 \zeta_2}}{\zeta_1 + \zeta_2}. \tag{2.9}$$

The matched-medium possibility is confirmed, and it is straightforward to verify the conservation relation

$$R^2 + T^2 = 1. \tag{2.10}$$

In section 4.2 of Mei's book [15] the analysis for long waves over a depth discontinuity is presented, using both the shallow water equations as well as linear potential theory. Chapter 2 of Dingemans book [5] also has a detailed presentation for wave reflection at a step, studied through the linear potential theory equations. Other examples are presented.

2.2 Reflection with the Presence of a Layer

As soon as we add another interface, namely a layer of a third material, this problem becomes much more complicated. The transmitted wave from the first interface will be an incoming wave at the second interface from the left. The reflected wave at the second interface will be an incoming wave from the right at the first interface. In this case we have a nonzero B_2 as a boundary condition at the first interface. Since we are not considering any dissipation there will be a, gradually decaying, signal reverberating forever within the layer of width L. We have that

$$\begin{bmatrix} A_2(0, t) \\ B_2(0, t) \end{bmatrix} = P_1 \begin{bmatrix} f(t) \\ B_1(0, t) \end{bmatrix}, \tag{2.11}$$

and

$$
\begin{bmatrix} A_3(L, t) \\ B_3(L, t) \end{bmatrix} = P_2 \begin{bmatrix} A_2(0, (t - L/c_2)) \\ 0 \end{bmatrix}.
\tag{2.12}
$$

Note that L/c_2 is the travel time within the layer from interface 1–2 up to interface 2–3. Within this layer region A_2 is invariant, namely a right-going disturbance. It will reflect at interface 2–3 and generate an incoming mode B_2 at interface 1–2. This back-and-forth reverberation ("ringing") leads to a complicated (mathematical) bookkeeping, not to mention when the total number of layers is large.

For the reasons above people have considered what is known as the Goupillaud medium. Goupillaud refers to a layered model where the travel time over each layer is the same, therefore the delay component is constant and is better dealt with in Fourier space as a phase factor. The frequency dependent propagator, from medium 1 up to medium 3, can be written in the form ([11, p. 43])

$$
\begin{bmatrix} \widehat{A}_3(\omega) \\ 0 \end{bmatrix} = \widehat{P} \begin{bmatrix} \widehat{f}(\omega) \\ \widehat{B}_1(\omega) \end{bmatrix},
\tag{2.13}
$$

where $\widehat{P} = \widehat{P}_2 \widehat{P}_1$ and

$$
\widehat{P}_1 = \begin{bmatrix} p^+ & p^- \\ p^- & p^+ \end{bmatrix}, \quad \widehat{P}_2 = \begin{bmatrix} p^+ e^{i\omega L/c_2} & p^- e^{-i\omega L/c_2} \\ p^- e^{-i\omega L/c_2} & p^+ e^{i\omega L/c_2} \end{bmatrix}.
$$

We have that $\widehat{B}_1(\omega) = \widehat{R}(\omega)\widehat{f}(\omega)$ and $\widehat{A}_3(\omega) = \widehat{T}(\omega)\widehat{f}(\omega)$. The reflection and transmission coefficients for the layer can be expressed as

$$
\widehat{R}(\omega) = \frac{\widehat{R}_2 e^{2i\omega L/c_2} + \widehat{R}_1}{1 + \widehat{R}_1 \widehat{R}_2 e^{2i\omega L/c_2}}, \quad \widehat{T}(\omega) = \frac{\widehat{T}_1 \widehat{T}_2 e^{i\omega L/c_2} + \widehat{R}_1}{1 + \widehat{R}_1 \widehat{R}_2 e^{2i\omega L/c_2}},
\tag{2.14}
$$

where as expected, $\widehat{R}_j$ and $\widehat{T}_j$, $j = 1, 2$ are given as in (2.9).

The simplest shallow water equivalent of a layer is system (2.4), where the depth $h(x)$ is piecewise constant, taking on three different values: h_1, h_2 and h_3. The region that plays the role of a layer has depth h_2. We have depicted in Fig. 2 a shallow water channel with steps that correspond to four layers in the associated acoustic model. The layer depths are respectively equal to 0.15, 0.60, 0.25 and 0.675.

2.3 Wave Reflection in a Finely Layered Medium

Acoustic waves provide a reasonable model for seismic waves through the crust of the Earth in oil exploration. The Earth crust is modeled as a finely layered medium

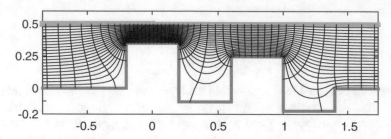

Fig. 2 Shallow water channel with a geometry equivalent of having four layers; four intermediate regions of constant speed (constant depth). The level curves depicted represent a curvilinear coordinate system which arises from the conformal mapping of a uniform flat strip onto the physical (polygonal shaped) domain

[11]. A good mathematical setting to start studying this nontrivial problem is the 1D model for pulses traveling over thousands of heterogenous layers. As shown earlier, our introductory problem, very quickly, becomes quite nontrivial in the presence of multiple layers, multiple reflectors. The formulation can be in terms of matrix products [11], as we have indicated in the single layer problem, or more generally using the alternative (and more general approach) based on differential equations for the propagator matrix. This eventually leads to random ordinary differential equations (ODEs) and probabilistic results, along the lines of central limit theorems [11], regarding expectations of the reflected or transmitted signals.

In section 4.1.1 of the book by Fouque et al. [11] a nice deterministic example is presented. It considers a finely layered periodic medium with alternating acoustic speeds c_a and c_b, together with the matched medium condition through the impedances: $\zeta_a = \zeta_b$. The medium is assumed to have $2N$ layers. The propagator matrix product is examined in the limit as $N \to \infty$ and the layer spacing going to zero, keeping the multilayer slab fixed in $[0, L]$. As a result one obtains the effective pulse speed $\bar{c}$ in the homogenized (composite) medium, which is built of two materials a and b. The effective speed is the harmonic average of the individual component speeds:

$$\bar{c} = \left[\frac{1}{2} \left(\frac{1}{c_a} + \frac{1}{c_b} \right) \right]^{-1}. \tag{2.15}$$

In this hyperbolic system of equations the effective speed turns out to be frequency independent. A pulse, which is broad with respect to the fine layers, propagates as a traveling wave. The matched impedance condition tells us that no reflection is observed at numerous interfaces. Hence, asymptotically speaking, the pulse propagates in an effectively homogeneous medium. Nevertheless the pulse has an effective speed different from a simple arithmetic average of the underlying material-speeds. Similar homogenized behavior is observed for water waves over rapidly varying periodic topographies, even in the nonlinear regime [24]. This has been illustrated computationally in some articles, as for example [1, 21]. It is a very good benchmarking exercise for numerical methods that are designed for multi-scale problems of waves propagating over a microstructure for large distances.

In section 4.5.1 [11] the study is carried over to a finely layered random medium, with a constant impedance. The propagation speeds at each layer are independent identically distributed, (strictly) positive random variables C_j. The homogenization calculation follows the lines of the Law of Large Numbers. In the limit as the (constant) layer width $l \to 0$, the effective speed is calculated so that the propagator becomes the identity matrix. The effective (or homogenized) propagation speed is expressed as

$$\bar{c} = \left(\mathbb{E}\left[\frac{1}{C_1} \right] \right)^{-1}. \tag{2.16}$$

This is also observed for long (linear) water waves propagating in a channel with randomly varying steps. Long nonlinear water waves have also been studied in the presence of a periodic and a random microstructure, regarding depth variations. This will be further discussed in Sect. 4.

3 Reduced Modeling

The mathematical exercise of using system (2.4) with a non-smooth topography follows from that of the acoustic model. It is a good warmup problem in order to consider a layered topography. Nevertheless in the presence of a rapidly varying (non-smooth) topography, the shallow water system is not asymptotically valid as a reduced model arising from the potential theory equations [13]. We will outline the reasons why this is so.

Consider the potential theory formulation arising from Euler's equations [26]. Let variables with physical dimensions be denoted with a tilde. We introduce the length scales σ (a typical pulse width or wavelength), h_0 (a typical depth), a (a typical wave amplitude), l_b (the horizontal length scale for bottom irregularities) and L (the total length of the rough region). The acceleration due to gravity is denoted by g and the reference shallow water speed is $c_0 = \sqrt{g h_0}$. Dimensionless variables are then defined in a standard fashion [24, 26] by having

$$\tilde{x} = \sigma x, \qquad \tilde{y} = h_0\, y, \qquad \tilde{t} = \left(\frac{\sigma}{c_0} \right) t,$$

$$\tilde{\eta} = a\, \eta, \qquad \tilde{\phi} = \left(\frac{g \sigma a}{c_0} \right) \phi, \qquad \tilde{h} = h_0\, H\left(\frac{\tilde{x}}{l_b} \right).$$

The velocity potential $\phi(x, y, t)$ and wave elevation $\eta(x, t)$ satisfy the dimension-less equations [26]:

$$\beta\,\phi_{xx} + \phi_{yy} = 0, \quad \text{for } -H(x/\gamma) < y < \alpha\eta(x, t),$$

with the nonlinear free surface conditions

$$\eta_t + \alpha\phi_x\eta_x - \frac{1}{\beta}\phi_y = 0,$$

$$\eta + \phi_t + \frac{\alpha}{2}\left(\phi_x^2 + \frac{1}{\beta}\phi_y^2\right) = 0,$$

at $y = \alpha\eta(x, t)$. The Neumann condition at the impermeable bottom is

$$\phi_y + \frac{\beta}{\gamma}\,H'(x/\gamma)\phi_x = 0.$$

The bottom topography is described by $y = -H(x/\gamma)$ where

$$H(x/\gamma) = \begin{cases} 1 + n(x/\gamma), & \text{when } 0 < x < L, \\ 1, & \text{when } x \leq 0 \text{ or } x \geq L. \end{cases}$$

The following dimensionless parameters arise: $\alpha = a/h_0$ (the nonlinearity parameter), $\beta = h_0^2/\sigma^2$ (the dispersion parameter) and $\gamma = l_b/\sigma$ (how rapidly varying bottom irregularities are compared to the wave-scale). When $\gamma \ll 1$ the bottom topography is denoted by the rapidly varying function $-n(x/\gamma)$. The topography can be of large amplitude about the flat bottom at depth equal to one. As depicted in Fig. 2 the function $n(x/\gamma)$ does not have to be smooth.

Let all three parameters be $O(\varepsilon)$ indicating that we are setting ourselves in the weakly nonlinear, weakly dispersive (long wave) regimes, in the presence of a rapidly varying topography. Hamilton [13] discusses that in this regime the formal asymptotic analysis to deduce reduced Boussinesq (or shallow water) models does not make sense. Hamilton's presentation does not use dimensionless variables. The asymptotic expansion about the flat bottom can be repeated with the dimensionless equations as found in Nachbin [20]. Performing Hamilton's calculations for the rapidly varying topography one will find $O(1/\varepsilon)$ terms that arise from derivatives of the bottom profile $H(x/\varepsilon)$. These will eventually reorder terms in the formal expansion and destroy any hope for convergence. Hence the shallow water equations in the presence of rapidly varying topographies is not valid as an asymptotic approximation of the potential theory equations. An outline of Hamilton's calculation [13] is given below.

For this reason, in analyzing long linear ($\alpha = 0$) waves over rapidly varying random topographies, Nachbin and Papanicolaou [23] considered the linear (full) potential theory equations. The probabilistic modeling, through random ordinary differential equations (ODEs), is presented in [11, 23] and the references within. Additional discussion and results are given in Nachbin [21], where a very simple example shows why ensemble averaging has to be performed in a random moving frame. The example consists of the unidirectional wave equation ($u_t + C\,u_x = 0$) with a random (constant) speed C. Suppose we are uncertain about the exact propagation speed of a wave and we add uncertainty to the model through (for example) a normally distributed C. Ensemble averaging in a fixed reference frame leads to a spurious attenuation of the wave. This is shown in detail in [21].

The ensemble averaging for the linear potential theory formulation in [23] is based on a central limit theorem framework developed by George Papanicolaou and collaborators [11]. As a result one has that (statistically speaking) the reflection process governed by the linear potential is the same as that of the acoustic system [21]. Therefore it is the same as for the shallow water (long wave) model, even though the asymptotic model reduction mentioned above was not valid.

But in Nachbin [20] a valid asymptotic (long wave) model was produced. In particular a model that is of interest for nonlinear (solitary-type) waves interacting with highly disordered bottom variations. This valid Boussinesq model was achieved by first performing the conformal mapping from a canonical flat strip onto the undisturbed corrugated physical domain. The mapping amounts to being the same as using an orthogonal curvilinear coordinate system in the physical domain, as shown in Fig. 2. A very convenient property concerns the Laplacian which is invariant under this change of coordinates which is boundary fitted [23]. The corrugated bottom is a level curve of the curvilinear coordinate system which trivializes the Neumann condition there. This avoids many derivatives of the bottom profile and therefore the formal asymptotic analysis targeting a reduced Boussinesq (long wave) model is much better behaved in the curvilinear coordinate system. This is true even if the topography is rapidly varying and/or has a non-smooth bottom [22]. This change of formulation has a positive numerical impact. In [18] it was shown that the conformal mapping pre-conditions the model for numerical discretization.

To put all these comments into perspective we outline some calculations in the following subsection.

3.1 The Conventional Depth Expansion

With the linear potential theory equations we better illustrate some shortcomings when using the conventional power series expansion in the presence of rapidly varying bottom topographies. The nonlinear case will be discussed subsequently in the curvilinear coordinates framework.

The linear potential theory equations are obtained with $\alpha = 0$. In the long wave regime, propagating over rapidly varying depth variations, we will set for example $\beta = \gamma = \varepsilon$. We have the wide-pulse-microstruture interaction regime which allows the study of solitary waves over rapidly varying random topography. Hamilton [13] did not consider a dimensionless system. Hence as a first exercise, lets consider

$\beta = \gamma = 1$. In the interest of a future comment we will keep the parameter γ in the equations, even though it is equal to one.

The linear equations are solved in the undisturbed fluid domain, with

$$\phi_{xx} + \phi_{yy} = 0 \quad \text{for} \quad -H(x/\gamma) < y < 0,$$

together with the linear "free surface" condition at the $y = 0$,

$$\phi_{tt} + \phi_y = 0,$$

where the free surface elevation η has been (momentarily) removed from the problem. It is recovered at the end by using $\eta = -\phi_t$. The Neumann condition at the impermeable bottom is

$$\phi_y + (1/\gamma) \, H'(x/\gamma)\phi_x = 0.$$

The conventional Taylor series expansion near the bottom [13, 26] has an ansatz of the form

$$\phi(x, y, t) = f^{(0)}(x, t) + (y + H(x/\gamma)) \, f^{(1)}(x, t) + \frac{1}{2}(y + H(x/\gamma))^2 \, f^{(2)}(x, t) + \dots$$

We should keep in mind that at first $\gamma = 1$. By collecting terms of equal powers in $(y + H(x))$, one gets the recurrence relation [13]

$$f^{(m+2)} = -\left\{ f_{xx}^{(m)} + 2H' \, f^{(m+1)} + H'' \, f^{(m+1)} \right\} / (1 + H'^2). \tag{3.1}$$

The Neumann condition yields

$$f^{(1)}(x, t) = -\frac{H' f_x^{(0)}}{1 + H'^2}. \tag{3.2}$$

For the flat bottom case, as presented in Whitham [26], $H' \equiv 0$ as well as all odd terms $f^{(2m+1)}(x, t)$, $m = 0, 1, 2, \dots$. In the flat bottom case one can operate only with the order-zero reduced potential $f_0(x, t)$ and at the end the zero subscript is dropped. Expression (3.1) allows for the high order reduced potential terms to be formally incorporated, as needed [13].

As discussed in Hamilton [13] (top of page 292), for the long wave case $(\beta = O(\varepsilon))$ in the presence of rapidly varying bottom variations $(H(x/\gamma); \; \gamma \ll 1)$ the asymptotics will break down. Derivatives of the bottom profile give rise to powers of $O(1/\varepsilon)$ that will eventually promote a reordering in the formal asymptotic expansion. Hamilton indicates that the errors for truncating the series expansion are no longer (formally) decaying.

In the need for an asymptotic model which can accommodate rapidly varying layered bottom topographies, Nachbin [20] revisited the work of Hamilton. The goal was to obtain a valid wave model amenable to extending several results from the linear hyperbolic (acoustic) waves case to the weakly nonlinear, weakly dispersive (water) waves case. This led to considering Hamilton's conformal mapping framework but using dimensionless variables to have better control on several underlying regimes. The idea is to have a canonical domain in the form of a uniform strip. The mapping takes the canonical domain onto the corrugated physical domain. From a slightly different perspective, ignoring the canonical domain, the conformal mapping amounts to using a curvilinear change of coordinate system which is orthogonal and boundary fitted, as depicted in Fig. 2.

3.2   The Conformal Mapping Formulation

We briefly review the conformal mapping framework, which can be found in detail in references [6, 20, 22]. Define a conformal mapping F from the (scaled) canonical domain in the complex w-plane onto the physical domain in the complex Z-plane [20]. The notation is such that $w = \xi + i\zeta$ while $Z = F(w) = x(\xi, \zeta) + iy(\xi, \zeta)$. In the curvilinear coordinate system (ξ, ζ) the nonlinear potential theory equations are given as

$$\phi_{\xi\xi} + \phi_{\zeta\zeta} = 0, \quad -\sqrt{\beta} < \zeta < \alpha\sqrt{\beta}N(\xi, t), \tag{3.3}$$

together with the kinematic and Bernoulli boundary conditions

$$|J|N_t + \phi_\xi\, N_\xi - \phi_\zeta = 0, \tag{3.4}$$

$$|J|\,(g\eta + \phi_t) + \frac{1}{2}|\nabla\phi|^2 = 0, \tag{3.5}$$

along the free surface's pre-image $\zeta = \alpha\sqrt{\beta}N(\xi, t)$. The trivial Neumann condition

$$\phi_\zeta = 0, \tag{3.6}$$

if defined at the bottom $\quad \zeta = -\sqrt{\beta}$ of the shallow channel. The Jacobian of the change of coordinates is denoted by $|J|(\xi, \zeta)$ and ∇ is taken in the $\xi\zeta$ variables. Under the mapping F we have that $x + i\eta(x, t) = F(\xi + iN(\xi, t))$, which establishes a functional relation between the two free surface representations η and N. Details of the numerical conformal mapping formulation can be found in Fokas and Nachbin [6]. Using the Cauchy-Riemann equations we have that $|J|(\xi, t) = y_\xi^2 + y_\zeta^2$. In the weakly nonlinear regime the Jacobian is well approximated [6, 20, 22] by a time independent metric coefficient M, where $|J| \approx M^2(\xi)$, and $M(\xi) \equiv$

$y_\zeta(\xi, 0) = 1 + m(\xi)$, with

$$m(\xi; \sqrt{\beta}, \gamma) \equiv \frac{\pi}{4\sqrt{\beta}} \int_{-\infty}^{\infty} \frac{n(x(\xi_0, -\sqrt{\beta})/\gamma)}{\cosh^2 \frac{\pi}{2\sqrt{\beta}}(\xi_0 - \xi)} d\xi_0. \tag{3.7}$$

The asymptotic analysis in [20] yields, what we have called, the terrain-following Boussinesq system:

$$M(\xi)\eta_t + [(1 + \alpha\eta/M(\xi)) U]_\xi = 0,$$

$$U_t + \eta_\xi + \tfrac{\alpha}{2}\left[U^2/M^2(\xi)\right]_\xi - \tfrac{\beta}{3}U_{\xi\xi t} = 0. \tag{3.8}$$

Here U is the depth-averaged horizontal velocity. Other (asymptotically equivalent) reduced models can be obtained with improved dispersion properties [19]. The classical shallow water regime eliminates all integer-order terms in β, which is the same as setting $\beta = 0$. With this conformal mapping setting we can take the long wave regime by eliminating all terms of $O(\beta)$ and above, but through the metric term M, keeping an $O(\sqrt{\beta})$ mollifying effect [22] which smooths out bottom discontinuities, as well as averages rapidly varying features [6, 22]. Thus the valid underlying shallow water system is

$$\eta_t + \tfrac{1}{M(\xi)} [(1 + \alpha\eta/M(\xi)) U]_\xi = 0,$$

$$U_t + \eta_\xi + \tfrac{\alpha}{2}\left[U^2/M^2(\xi)\right]_\xi = 0. \tag{3.9}$$

By the conformal mapping we are assured that $M(\xi)$ is C^∞ and strictly positive. This is true in the case that the topography has corners and also in the case when the topography has large rapid oscillations about its mean level. The linear ($\alpha = 0$) analogue for the acoustic equations in a layered medium (such as in Fig. 2) is

$$\eta_t + \tfrac{1}{M(\xi)}U_\xi = 0,$$

$$U_t + \eta_\xi = 0. \tag{3.10}$$

4 More Advanced Developments

In order to guide the interested reader through some recent research, we briefly comment on some references. The description will be informal, avoiding equations, attempting to stimulate further studies. In the interest of space it will be more focused on the author's work, keeping in mind that many other references can be found in the articles here cited, pointing to many important work but other authors.

4.1 Regarding Wave Reflection

In the acoustic waves literature there is a very interesting phenomenon called by mathematicians as the O'Doherty-Anstey (ODA) approximation. An acoustic pulse propagating in a random, rapidly varying, layered medium displays an *apparent attenuation* due to the multiple scattering that takes place at the interface of the numerous layers. The mathematical theory starts with the variable coefficient, hyperbolic system presented in the present paper. In a proper scaling regime, to leading order, the transmitted pulse-shaped pressure wave is given by the convolution of its initial profile with a Gaussian kernel. The asymptotic analysis is framed as a near wavefront approximation. The Gaussian kernel can be viewed as a Gaussian low pass filter, as well as the heat kernel for a diffusion equation. Note that the Gaussian filter, which also characterizes the *apparent diffusion*, indicates wave energy being converted from the wavefront to its forward and backward (disordered) scattered signals: the pulse's coda and the reflected wave. We have a conservation law, so no energy is being lost through this *apparent diffusion*. Indeed it is being extracted from the wavefront where the asymptotic approximation is being made. As mathematical references, we mention the probabilistic formulation by Clouet and Fouque [4] and the deterministic formulation by Berlyand and Burridge [3]. The ODA approximation is also called as the *pulse-spreading formula*, as presented in chapter 8 of the book by Fouque et al. [11]. A higher dimension analysis for the ODA approximation is presented by Sølna and Papanicolaou [25] where ray theory takes care of slowly varying features of the 2D medium.

In Nachbin [20] the weakly nonlinear weakly dispersive water wave model (3.8) was deduced having in mind rapidly varying random topographies and the ODA theory developed for acoustic waves. By setting $\alpha = 0$ and $\beta = \varepsilon$, Eq. (3.8) become a dispersive perturbation of the linear acoustic model. The theory presented by Berlyand and Burridge [3] was extended to weakly dispersive waves. It used the Riemann Invariants of the underlying hyperbolic system, in the dispersive setting where characteristics are no longer present and these invariants are no longer invariant, but are slowly varying. Nevertheless, the Riemann Invariants suggest a very useful change of variables and the dispersive pulse-shaping formula was obtained. It comes out in the form of a Fourier integral and was compared, with great accuracy, to numerical simulations resolving all scales. These theoretical and numerical results are given in Muñoz and Nachbin [17]. A graphical example of the apparent diffusion is displayed in Fig. 3. The theory formulated [17] has its Fourier transform in terms of the delay time $\tau \equiv t - x$, where t is the usual time variable and x is travel time over the distance ξ, where $x \equiv \int_0^\xi (1/C(s))ds$. The reference speed is normalized to 1 and therefore there is no delay in the absence of inhomogeneities. This is observed in the upper part of Fig. 3 where the (dashed) initial Gaussian-shaped pulse is compared with the final pulse profile. Under weak dispersion we see a short oscillatory (dispersive) coda and no delay. In the bottom graph of Fig. 3 we compare a well resolved numerical simulation (solid line) with the theoretical Fourier integral. We observe a minor delay, at the pulse's peak, and the

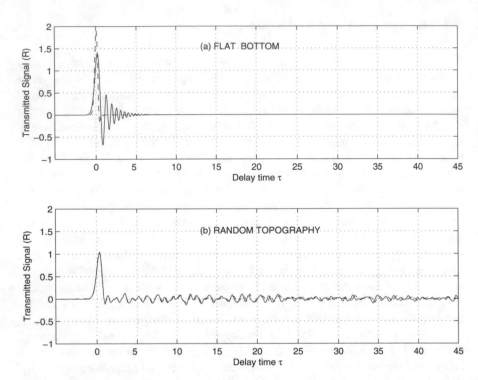

Fig. 3 Top: the initial pulse profile (dashed line) and the propagated pulse in a flat bottom configuration (solid line). Bottom: the numerical solution (solid line) and the theoretical Fourier integral representation (dashed line)

apparent diffusion through the broadening and attenuation of the pulse. The energy of the leading wavefront is converted, through multiple scattering, into the long disordered coda behind the front. To leading order the Fourier integral represents a convolution with a Gaussian kernel, analogous to a heat kernel. Subsequently in [8] the probabilistic formulation was presented, leading to more universal results regarding the class of topographic disorder. For example, the fact that the pulse-shaping is very stable regarding different realizations of the random medium.

Regarding nonlinear waves, Fouque et al. [9] considered the hyperbolic system for shallow water waves with a randomly varying depth. In the absence of bathymetric variations a shock eventually forms. Nevertheless in the presence of a disordered medium, in the regime of the apparent diffusion, a viscous shock is observed. The theory considers the Riemann Invariants for the underlying conservation law. In the presence of multiple scattering they are not invariant and exchange energy in a similar fashion to modes A and B discussed earlier. The asymptotic analysis for the right propagating mode A leads to the following result: up to a random shift, mode A is effectively governed by a viscous Burgers equation, where the apparent viscosity is related to an integral of the autocorrelation function of the random depth fluctuations.

This led us to exploring with solitary waves in a rapidly varying disordered medium. By using again the underlying Riemann Invariants, now within the weakly nonlinear, weakly dispersive Boussinesq system, Garnier et al. [12] deduced an effective Burgers-KdV (Korteweg-de Vries) equation for the right propagating mode. The theoretical results were qualitatively in very good agreement with numerical simulations.

Another problem of great interest relates to waveform inversion through the time-reversal of the reflected signal [11]. The author has worked on the one-dimensional problem where one can consider time-reversal in reflection or time-reversal in transmission. In the direct problem a pulse-shaped wave is sent onto the random half-space, say placed along the positive axis, incoming from the left. The reflected signal is recorded somewhere near the beginning of the random medium. This is displayed schematically in Fig. 4. Here the initial pulse is in the form of the derivative of a Gaussian. As before the pulse is under the apparent diffusion as it propagates to the right and has its energy converted to the long multiply-scattered signal to its left. The disordered profile over the interval $\xi \in [0, 45]$ is a snapshot of the superposition of a left and right-going wave. On the other hand the disordered signal on the negative axis is a left-going (reflected) wave propagating over a flat bottom. This signal (singled out at the top) is recorded, time-reversed and sent back into the variable depth region to recompress, under reflection, and generate

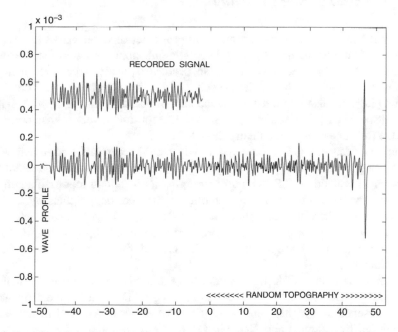

Fig. 4 Snapshot of a pulse propagating over a random topography, positioned along the positive axis $\xi > 0$. In the detail above, is displayed the recorded reflected signal that, through time-reversal, will refocus into a derivative of a Gaussian when sent back into the random medium

a (reduced) copy of the initial pulse. Hence a waveform inversion with a fraction of the energy. Having recorded the reflected signal for a long but finite amount of time amounts to keeping only a fraction of the of energy. Time-reversal is, in a sense, a play-back process where "last-information-out" (as an echo of the medium) will be the "first-information-in" into the same random medium, in the next step. In a hyperbolic system the data is moved from one characteristic to the other. The disordered reflected signal that was moving to the left now is a signal moving (backwards) to right. This new initial data will impinge on the same random slab under the same system of partial differential equations. It is remarkable that the random reflected signal of this time-reversed data refocuses, within the random medium where it was generated, giving rise to a smooth pulse with the same shape of the original data of the direct case. This pulse is a reduced copy because we used only part of the energy. In water waves this is a mechanism for the waveform inversion of, for example, a tsunami.

In papers [8, 18, 19] time-reversal of one-dimensional (1D) linear water waves were studied. The time-reversed refocusing of two-dimensional (2D) linear shallow water waves was performed by Fouque et al. [10]. Nonlinear wave refocusing was then presented for the shallow water model in [9] and for solitary waves in [7].

4.2   Regarding Wave Modeling

Much has been outlined above for effective properties of water waves in a rapidly varying medium. In particular, many interesting phenomena arise related to wave scattering. As mentioned, studying these problems greatly depended on having good reduced models at hand. Here we should mentioned not only the Boussinesq system (3.8), but also improved ones such as those describe in [19], which display better dispersive properties and are more accurate for time-reversal when compared with the full (linear) potential theory model.

Mei and Hancock [16] considered slowly modulated weakly nonlinear waves propagating over a random topography. They obtained an effective nonlinear Schrödinger equation (NLS) with a damping term containing topographic effects. This NLS was generalized to large-amplitude variable topographies by Luz and Nachbin [14], where the topography amplitude affected the focusing/defocusing critical wavenumber. The large-amplitude variable random topography regularizes the problem in the sense of pushing the critical focusing point to a higher wavenumber.

Recently the author, with collaborators, has explored water wave models in three-dimensions. Namely the 3D fluid problem having a 2D surface wave propagating in the presence of a highly variable topography. A linear potential theory model is proposed by Andrade and Nachbin [1] where a Dirichlet-to-Neumann (DtN) operator is constructed in the presence of highly variable topography. The DtN operator analytically reduces one dimension of the problem. It leads to a Fourier integral, computed through an FFT, which encodes information of the harmonic velocity potential and the vertical structure of the flow. This method worked very well for the linear problem but is expensive. In particular for 3D nonlinear problems the DtN operator has to be updated continuously and displays numerical instabilities due to the deformed varying free surface. Alternatively, Andrade and Nachbin [2] deduced a 2D Boussineq system which generalizes that presented in [20]. This reduced model starts from the 3D Laplace equation in a domain with ridge-like (1D) topography. Conformal mapping is used to flatten out the topography, but in this (3D) case the Laplacian is no longer invariant. Nevertheless the asymptotic model reduction goes through and a variable coefficient Boussinesq system is obtained and tested numerically.

Acknowledgements The author thanks the referee for the careful reading and suggestions. The author is grateful to the Erwin Schrödinger International Institute for Mathematics and Physics, for the support and hospitality during the *Nonlinear Water Waves: An Interdisciplinary Interface* workshop. The work of A.N. was supported in part by: CNPq under (PQ-1B) 454027/2008-7; FAPERJ project #112112/08 and also FAPERJ *Cientistas do Nosso Estado* project #102793/08.

References

1. D. Andrade, A. Nachbin, A three-dimensional Dirichlet-to-Neumann operator for water waves over topography. J. Fluid Mech. **845**, 321–345 (2018)
2. D. Andrade, A. Nachbin, Two-dimensional surface wave propagation over arbitrary ridge-like topographies. SIAM J. Appl. Math. **78**(5), 2465–2490 (2018)
3. L. Berlyand, R. Burridge, The accuracy of the O'Doherty-Anstey approximation for wave propagation in highly disordered stratified media. Wave Motion **21**, 357–373 (1994)
4. J.F. Clouet, J.P. Fouque, Spreading of a pulse traveling in a random media. Ann. Appl. Prob. **4**, 1083–1097 (1994)
5. M.W. Dingemans, *Water Wave Propagation Over Uneven Bottoms* (World Scientific, Singapore, 1997)
6. A.S. Fokas, A. Nachbin, Water waves over a variable bottom: a non-local formulation and conformal mappings. J. Fluid Mech. **695**, 288–309 (2012)
7. J.P. Fouque, J. Garnier, J.C. Muñoz Grajales, A. Nachbin, Time reversing solitary waves. Phys. Rev. Lett. **92**(9), 094502-1 (2004)
8. J.P. Fouque, J. Garnier, A. Nachbin, Time reversal for dispersive waves in random media. SIAM J. Appl. Math. **64**, 1810–1838 (2004)
9. J.P. Fouque, J. Garnier, A. Nachbin, Shock structure due to stochastic forcing and the time reversal of nonlinear waves. Phys. D **195**, 324–346 (2004)
10. J.P. Fouque, J. Garnier, A. Nachbin, K. Sølna, Time-reversed refocusing for point source in randomly layered media. Wave Motion **42**, 238–260 (2005)
11. J.P. Fouque, J. Garnier, G.C. Papanicolaou, K. Sølna, *Wave Propagation and Time Reversal in Randomly Layered Media* (Springer, Berlin, 2007)
12. J. Garnier, J.C. Muñoz Grajales, A. Nachbin, Effective behavior of solitary waves over random topography. Multiscale Model. Simul. **6**, 995–1025 (2007)
13. J. Hamilton, Differential equations for long-period gravity waves on a fluid of rapidly varying depth. J. Fluid Mech. **83**, 289–310 (1977)
14. A.M.S. Luz, A Nachbin, Wave packet defocusing due to a highly disordered bathymetry. Stud. Appl. Math. **130**, 393–416 (2013)
15. C.C. Mei, *Applied Dynamics of Ocean Surface Waves* (World Scientific, Singapore, 1989)

16. C.C. Mei, M.J. Hancock, Weakly nonlinear surface waves over a random seabed. J. Fluid Mech. **475**, 247–268 (2003)
17. J.C. Muñoz-Grajales, A. Nachbin, Dispersive wave attenuation due to orographic forcing. SIAM J. Appl. Math. **64**, 977–1001 (2004)
18. J.C. Muñoz-Grajales, A. Nachbin, Stiff microscale forcing and solitary wave refocusing. Multiscale Model. Simul. **3**, 680–705 (2005)
19. J.C. Muñoz-Grajales, A. Nachbin, Improved Boussinesq-type equations for highly-variable depths. IMA J. Appl. Math. **71**, 600–633 (2006)
20. A. Nachbin, A terrain-following Boussinesq system. SIAM J. Appl. Math. **63**, 905–922 (2003)
21. A. Nachbin, Discrete and continuous random water wave dynamics. Discrete Contin. Dyn. Syst. A **28**, 1603–1633 (2010)
22. A. Nachbin, Conformal mapping and complex topographies. *Lectures on the Theory of Water Waves*, ed. by T.J. Bridges M.D. Groves, D.P. Nicholls. Lecture Notes Series, vol. 426 (London Mathematical Society, London, 2015), pp. 203–225
23. A. Nachbin, G.C. Papanicolaou, Water waves in shallow channels of rapidly varying depth. J. Fluid Mech. **241**, 311–332 (1992)
24. R.R. Rosales, G.C. Papanicolaou, Gravity waves in a channel with a rough bottom. Stud. Appl. Math. **68**, 89–102 (1983)
25. K. Sølna, G.C. Papanicolaou, Ray theory for a locally layered medium. Waves Random Media **10**, 151–198 (2000)
26. G.B. Whitham, *Linear and Nonlinear Waves* (Wiley, London, 1974)

Global Diffeomorphism of the Lagrangian Flow-map for a Pollard-like Internal Water Wave

Mateusz Kluczek and Adrián Rodríguez-Sanjurjo

Abstract In this article we provide an overview of a rigorous justification of the global validity of the fluid motion described by a new exact and explicit solution prescribed in terms of Lagrangian variables of the nonlinear geophysical equations. More precisely, the three dimensional Lagrangian flow-map describing this exact and explicit solution is proven to be a global diffeomorphism from the labelling domain into the fluid domain. Then, the flow motion is shown to be dynamically possible.

Keywords Global diffeomorphism · Geophysical internal water waves · Exact and explicit solution

Mathematics Subject Classification (2000) 35A16, 35C05, 35Q86, 35Q35

1 Introduction

Exact and explicit solutions of the nonlinear geophysical equations are very rare. The use of the Lagrangian framework has however produced some remarkable results. In order to conclude that these solutions produce a valid fluid motion, they need to be subjected to a rigorous analysis beyond the confirmation that they satisfy the governing equations. In particular, we undertake for the first time such analysis to a three-dimensional exact solution derived for the internal water waves in a rotating system. It can be shown by explicit calculations that it is possible to have a motion of the whole fluid body where all particles describes circles with a radius dependent on the depth. It is important to emphasize that these exact and explicit solutions fulfill the governing equations locally in terms of Lagrangian labelling variables, but this does not take into full consideration a rigorous analysis

M. Kluczek (✉) · A. Rodríguez-Sanjurjo
School of Mathematical Sciences, University College Cork, Cork, Ireland
e-mail: m.kluczek@umail.ucc.ie; a.rodriguezsanjurjo@umail.ucc.ie

© Springer Nature Switzerland AG 2019
D. Henry et al. (eds.), *Nonlinear Water Waves*, Tutorials, Schools, and Workshops in the Mathematical Sciences, https://doi.org/10.1007/978-3-030-33536-6_2

of the global evolution of the fluid domain under wave propagation. We use degree-theoretic methods to illustrate the process and show that particles never collide and fill out the entire region where waves propagate, therefore the flow is said to be globally dynamically possible.

The solution analysed in this paper utilises the Lagrangian framework [1, 5] for providing an exact and explicit solution of the geophysical nonlinear water wave governing equations. Thus, it is included in a rare group of explicit solutions whose first example was the remarkable solution derived by Gerstner [11] and rediscovered in [12, 29, 30]. Gerstner's solution is the only-known exact solution to the nonlinear two-dimensional gravity wave problem under constant density. However, the fact that the resultant flow is rotational and cannot be generated by conservative forces has prevented the development of these type of solutions in favour of Stokes waves.

The idea behind Gerstner's wave is that the fluid motion is described by labelling individual particles moving in a roughly circular motion. The first Gerstner's wave generalisation was done by Pollard [28], extending Gerstner's wave for an incompressible vertically-stratified fluid in a rotating system and providing a genuinely three-dimensional water wave solution. The solution still describes circular particles paths but now the circular trajectories lie on a plane slightly tilted with respect to the local vertical. This solution, which is more adequate for flows outside the equatorial region, was generalised in [9] incorporating an underlying current allowing the solution to produce much more complex flows and including Pollard's and Gerstner's waves as particular cases. Subsequently, a new Pollard-like solution was derived to describe internal water waves [24].

All these solutions have in common that they are described in the Lagrangian framework. The validity of such construction where artificial labels are employed in order to prescribe the particles paths, relies on the fact that the flow map from the domain of the labels to the fluid domain is a global diffeomorphism. In this sense, the first rigorous mathematical analysis was accomplished in [3] for Gerstner's wave and subsequently in [13] by means of an elegant proof that inspired the subsequent studies of more general flows like equatorially-trapped water waves [34] and internal water waves [31]. Regarding Pollard's wave and its generalisation, they both were proven to be dynamically possible in [32]. It is the aim of this paper to complete the study of the Pollard-like internal wave solution [24] by providing a mathematical justification of the diffeomorphic character of the Lagrangian flow map describing it.

The straight forward implementation of Lagrangian solutions make them suitable for ocean simulation software and laboratory experiments. Studies like [26] has shown its applicability to ocean phenomena, whereas the oceanographical relevance of this solution is discussed in [2]. Consequently, there has been a significant increase of studies dealing with explicit solutions [3, 13, 35]. This remarkable solution was successfully extended to describe an extensive range of nonlinear geophysical water waves, for instance: weakly three-dimensional equatorially-trapped surface waves [6], internal waves [7, 8, 19], as well as edge waves [4, 20, 27]. In addition, Gerstner-like solutions may incorporate underlying currents [14, 23] providing a valuable insight into wave-current interactions (cf. [22] for the importance of the incorporation of underlying currents). Moreover, solutions to the geophysical equations incorporating centripetal forces and with a more accurate approximation of the gravitational force were derived in [15, 16]. The interested reader is referred to [17] for an extensive review of Gerstner-like solutions. Finally, although remarkable, these flows are not expected to be found in this form in the ocean; however, perturbations of such exact and explicit solutions can approximate real-life observations. In this sense, the hydrodynamical stability becomes a key factor (see [21] for a complete discussion of this issue).

2 Exact Pollard-like Internal Water Waves

2.1 The Governing Equations

In this section we provide a brief description of the governing equations representing nonlinear water waves followed by a section describing what it can be called a Pollard-like internal water wave solution of those equations. The solution is given for the f-plane equations characterizing internal water waves, which in this particular study specifies the oscillation of a thermocline (an interface separating two vertical ocean regions of constant density). In order to address the physical complexity of the stratification of the ocean, three layers are distinguished; a motionless abyssal deep-water region denoted by $\mathcal{S}(t)$ where the water density is given by the constant ρ_1 which is strictly greater than the density ρ_0 in the regions above it, a region denoted by $\mathcal{M}(t)$ within which the oscillations propagate and a near-surface layer $\mathcal{L}(t)$ where the motion is mainly due to the action of winds and where the oscillations of the thermocline can be seen as a small perturbation (see Fig. 1). Therefore, the geophysical internal water waves describing the oscillation of the thermocline propagate only in the layer $\mathcal{M}(t)$, with the amplitude decreasing as we ascend towards the surface. The amplitude of the waves decreases exponentially; moving upwards 1/10 of the wavelength, the amplitude of the wave is already 1/2 of the amplitude of the thermocline.

The internal water waves propagate zonally with a wavelength such that the effects of the Earth's rotation are significant. We take the Earth to be a perfect sphere of a radius $R = 6378$ km, rotating with a constant rotational speed $\Omega = 7.29 \times 10^{-5}$ rad s^{-1}. The flow is described by means of a rotating frame with the origin at a point on Earth's surface with the coordinates (x, y, z) representing the directions of longitude, latitude and local vertical, respectively. The governing equations for the internal geophysical ocean waves are given by the Euler equations incorporating the Coriolis terms accounting for Earth's rotation. Let (u, v, w) be the velocity field, ρ_0 the water density in the region $\mathcal{M}(t)$ where the internal waves

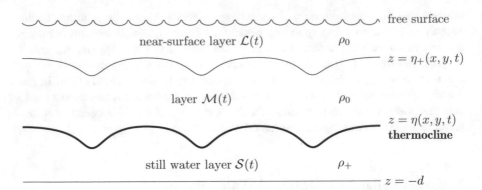

free surface

near-surface layer $\mathcal{L}(t)$ ρ_0

$z = \eta_+(x, y, t)$

layer $\mathcal{M}(t)$ ρ_0

$z = \eta(x, y, t)$
thermocline

still water layer $\mathcal{S}(t)$ ρ_+

$z = -d$

Fig. 1 The layers of the fluid domain at a fixed latitude y. The thermocline is described by a trochoid and separates a layer of less dense water overlaying a layer of more dense and colder water. The amplitude of the internal water wave decays exponentially and at a height of 1/10 of the wavelength above the thermocline the amplitude of these waves is less than 50% of its initial value at the thermocline

propagate, $g = 9.8\,\mathrm{m\,s}^{-2}$ the gravitational constant, and let P be the pressure. Then, the f-plane approximation of the Euler equations is given by

$$
\begin{cases}
u_t + uu_x + vu_y + wu_z + \hat{f}w - fv = -\dfrac{1}{\rho_0}P_x\,, \\[2mm]
v_t + uv_x + vv_y + wv_z + fu = -\dfrac{1}{\rho_0}P_y\,, \\[2mm]
w_t + uw_x + vw_y + ww_z - \hat{f}u = -\dfrac{1}{\rho_0}P_z - g\,.
\end{cases}
\tag{2.1a}
$$

where, if ϕ denotes the latitude, $f = 2\Omega\sin\phi$ and $\hat{f} = 2\Omega\cos\phi$ are the Coriolis parameters (see [10] for a detailed derivation of Eq. (2.1a)). In the f-plane approximation, by considering the flow propagating in a relatively narrow ocean strip in a small neighbourhood of fixed latitude, the Coriolis parameters can be taken as constants. In particular, for latitude of $45°$ North, the Coriolis parameters are $f = \hat{f} \approx 10^{-4}\,\mathrm{s}^{-1}$ while along the equator f vanishes and $\hat{f} = 2\Omega$. The Euler equations (2.1a) are coupled with the equation of mass conservation

$$
\nabla \cdot \vec{u} = 0\,,
\tag{2.1b}
$$

for an incompressible fluid such that

$$
\frac{D\rho}{Dt} = 0\,,
\tag{2.1c}
$$

where D/Dt represents the material derivative. Finally, boundary conditions must be introduced in order to obtain a valid model for the internal geophysical water waves. This results in the dynamic boundary condition

$$P = P_0 - \rho_+ gz \qquad \text{on the thermocline } z = \eta(x, y, t), \qquad (2.1d)$$

where P_0 is taken to be constant, and the kinematic boundary condition

$$w = \frac{D\eta}{Dt} \qquad \text{on the thermocline } z = \eta(x, y, t). \qquad (2.1e)$$

This last condition prevents mixing particles between the layers separated by the thermocline.

2.2 Internal Water Wave Solution

The new solution in terms of Lagrangian labelling variables [1] for the nonlinear internal water waves, derived by Kluczek in [24], is given by

$$\begin{cases} x = q - b\,e^{-ms}\sin[k(q - ct)], \\[2mm] y = r - d\,e^{-ms}\cos[k(q - ct)], \\[2mm] z = s - a\,e^{-ms}\cos[k(q - ct)], \end{cases} \qquad (2.2)$$

representing a travelling wave with a wave speed c. The Lagrangian labelling variables (q, r, s) belong to the set $\mathbb{R} \times (-r_0, r_0) \times (s_0, s_+)$. The parameter a corresponds to the amplitude of the wave, the parameter m is the so-called modified wavenumber and it is also a decay factor of the amplitude of the wave. The parameters b, d, m, c have to be suitably chosen in terms of a, f, and wavenumber k in order to provide a continuous pressure satisfying (2.1) and (2.2). In particular, we set $a > 0, k > 0$ and we require $m > 0$ in order to describe internal waves with amplitude decreasing with the height above the thermocline. The parameters b, d together with the parameter a are responsible for the shape of the closed particle trajectory.

The variables (q, r, s) do not represent initial positions but the centre of a circle made by each particle with a maximum radius of $1/m$. The paths of water particles are indeed circles due the non-zero vorticity of the Pollard-like solution (see [24] for a detailed derivation of the solution). Moreover, the orbits of the particles are tilted with an angle of $\arctan(-d/a)$ with respect to the local vertical and such that the inclination increases with the latitude (see Fig. 2). The meridional width is restricted by r_0 which expands about given latitude ϕ. In the flow domain the values s_0 and s_+ represent the thermocline and the upper boundary of the layer $\mathcal{M}(t)$ for

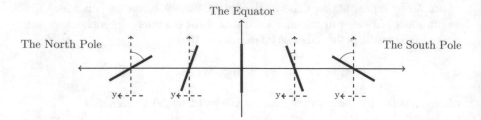

Fig. 2 Inclination of the particles' orbits as the latitude increases. At the Equator the orbit becomes vertical, as the parameter d is equal to zero, and as a result there is no motion of the particle in the meridional direction [24]

fixed latitude, respectively. We introduce only the relations between the mentioned parameters which are relevant to the proofs contained in this paper. The detailed analysis of solution (2.2) and the complete justification of these relations are given in [24].

The relations significant to our study are

$$b^2 = a^2 + d^2 \,, \tag{2.3}$$

and

$$am = bk \,. \tag{2.4}$$

The first relation follows from the boundary condition (2.1d). In order to satisfy the mass conservation, the solution (2.2) must preserve the volume, which holds if and only if the Jacobian determinant is independent of time. This is ensured by relation (2.4). Moreover, this Pollard-like solution provides the following relation between physical and Lagrangian labelling parameters

$$m^2 = \frac{k^4 c^2}{k^2 c^2 - f^2} \,,$$

and it follows that

$$m \geq k \,, \tag{2.5}$$

for the whole fluid domain. At the Equator $m = k$ and the solution (2.2) particularises to a Gerstner-like solution. Before proceeding to the main result establishing that (2.2) describes a dynamically possible fluid motion, we introduce a brief quantitative discussion.

Let us consider an internal water wave of a wavelength $L = 100$ m, which is associated with a wavenumber $k = 6.28 \times 10^{-2}\, \mathrm{m}^{-1}$. For waves propagating on latitude 45°N, where the density difference is taken to be $\Delta\rho/\rho = 6 \times 10^{-3}$, the

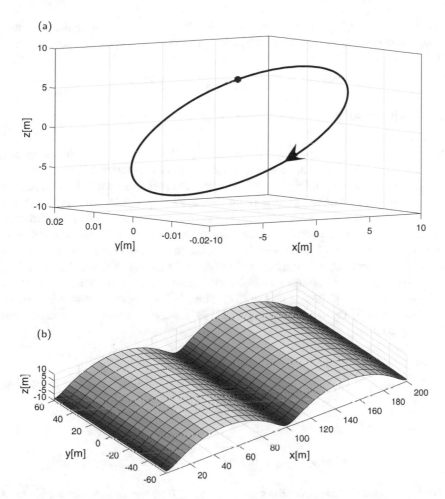

Fig. 3 Quantitative aspects of the internal water waves. (**a**) Sketch of a circular path of a water particle. (**b**) Internal water wave describing the oscillation of the thermocline

wave phase speeds derived from the model are $c \approx -0.9671\,\mathrm{m\,s^{-1}}$ for the westward propagating wave and $c \approx 0.9687\,\mathrm{m\,s^{-1}}$ for the eastward propagating wave. The value of the parameter m for such waves is respectively $m \approx 1.0341\,\mathrm{m^{-1}}$ and $m \approx 1.0323\,\mathrm{m^{-1}}$. The maximal amplitude of the internal water wave describing the oscillation of the thermocline is 15.91 m, whereas the amplitude of wave presented in Fig. 3 is approximately 8.4839 m, which is equivalent to the radius of the circular path traced by a water particle. The Coriolis parameters on latitude 45°N are $f = \hat{f} \approx 1.0309 \times 10^{-4}\,\mathrm{rad\,s^{-1}}$ and therefore the parameter d, which is responsible for the deviation of the particle path from the local vertical, is approximately $0.027\,\mathrm{s^{-1}}$. This accounts for an angle of the inclination of the circular particle path to the local vertical around $0.0016\,\mathrm{rad} = 0.094°$. Therefore, in this particular model for the wave presented in Fig. 3 at the highest and lowest point, representing the crest and trough of the wave, the distance of the particle from the local vertical plane is around 0.0139 m.

3 Main Results

Solutions given by a Lagrangian map like (2.2) not only need to satisfy the governing equations (2.1) but they also need to describe a motion of the whole fluid body, in this case the fluid region $\mathcal{M}(t)$, where the particles do not collide. A flow satisfying such conditions is said to be dynamically possible. The mathematical abstraction that encapsulates this idea is the requirement of the Lagrangian map to be a global diffeomorphism where the smoothness of the fluid quantities is preserved. In Theorem 3.2 we provide a proof of the diffeomorphic character of the map (2.2) that extends regularly to the boundary, rigorously showing that such flow is dynamically possible. Before addressing the main result, the problem is simplified following the same philosophy as in [32]. Thus, let us consider the special case of the solution (2.2) for $t = 0$ and express it as a mapping from the correspondent labelling domain, i.e.

$$(q, r, s) \in \mathscr{D} \longmapsto \mathscr{F}(q, r, s) = \begin{bmatrix} q - be^{-ms} \sin(kq) \\ r - de^{-ms} \cos(kq) \\ s - ae^{-ms} \cos(kq) \end{bmatrix}, \tag{3.1}$$

where

$$\mathscr{D} = \{(q, r, s) : q \in \mathbb{R}, \, r \in (-r_0, r_0), \, s \in (s_0, s_+)\}. \tag{3.2}$$

Then, it is possible to recover the general case $t \geq 0$ by the change of variables and a shift in the x component

$$\mathscr{F}(q - ct, s, r) + \begin{bmatrix} ct \\ 0 \\ 0 \end{bmatrix}.$$

Moreover, the y and z component of (3.1) are periodic in q with period $\lambda = 2\pi/k$, while the x component experiences a shift of λ. Therefore, if $\mathscr{F}$ is a global diffeomorphism from the domain

$$\mathscr{D}_\lambda = \{(q, r, s) : q \in (0, \lambda), \, r \in (-r_0, r_0), \, s \in (s_0, s_+)\}. \tag{3.3}$$

into its image so it will be $\mathscr{F}$ from the whole domain (3.2) into the interior of $\mathcal{M}(t)$. Finally, it will be also proven in Theorem 3.2 that the boundary of these domains are mapped into the boundary of the fluid region $\mathcal{M}(t)$. The following proposition establishes a local result regarding the map (3.1).

Proposition 3.1 *If*

$$1 - ame^{-ms_0} > 0, \tag{3.4}$$

then the map $\mathcal{F}$ is a local diffeomorphism from $\mathcal{D}_\lambda$ into $\mathcal{F}(\mathcal{D}_\lambda) \subset \mathcal{M}(t)$.

Proof The main idea behind this proof is to apply the Inverse Function Theorem. Hence, we start by obtaining the Jacobian matrix of $\mathcal{F}$

$$D\mathcal{F}_{(q,r,s)} = \begin{pmatrix} 1 - bke^{-ms}\cos(kq) & 0 & bme^{-ms}\sin(kq) \\ dke^{-ms}\sin(kq) & 1 & dme^{-ms}\cos(kq) \\ ake^{-ms}\sin(kq) & 0 & 1 + ame^{-ms}\cos(kq) \end{pmatrix}$$

and the corresponding Jacobian,

$$\begin{vmatrix} 1 - bke^{-ms}\cos(kq) & bme^{-ms}\sin(kq) \\ ake^{-ms}\sin(kq) & 1 + ame^{-ms}\cos(kq) \end{vmatrix} =$$

$$= 1 - abmke^{-2ms} + (am - bk)e^{-ms}\cos(kq).$$

From (2.4), the determinant is reduced to,

$$1 - a^2m^2e^{-2ms}.$$

Therefore, from the hypothesis (3.4), the Jacobian does not vanish for all $s \in (s_0, s_+)$. On the other hand, $\mathcal{F}$ has continuous partial derivatives in any neighbourhood contained in $\mathcal{D}_\lambda$. Hence, $\mathcal{F}$ is a continuously differentiable map such that its Jacobian is strictly positive, so $\mathcal{F}$ is a local diffeomorphism from $\mathcal{D}_\lambda$ into its range by the Inverse Function Theorem. □

In order to prove that $\mathcal{F}$ is a global diffeomorphism we will show that it is globally bijective. Let us start by the injectivity.

Proposition 3.2 *If*

$$1 - bme^{-ms_0} > 0, \tag{3.5}$$

then $\mathcal{F}$ is globally injective on $\overline{\mathcal{D}}$.

Proof Let us consider $\mathcal{F}$ expressed in the following form

$$\mathcal{F}(q, r, s) = (q, r, s) - \mathcal{G}(q, r, s),$$

where

$$\mathscr{G}(q, r, s) = e^{-ms}\Big(b \sin(kq),\, d \cos(kq),\, a \cos(kq) \Big).$$

We note now that $\mathscr{G}$ is a continuously differentiable map and that, for any convex domain, it satisfies the following mean-value-theorem type of inequality

$$|\mathscr{G}(q, r, s) - \mathscr{G}(\tilde{q}, \tilde{r}, \tilde{s})|_2 \le \max_{\tau \in [0,1]} \|D\mathscr{G}_{\tau(q,r,s)+(1-\tau)(\tilde{q},\tilde{r},\tilde{s})}\|_2 \cdot |(q, r, s) - (\tilde{q}, \tilde{r}, \tilde{s}))|_2$$

$$(3.6)$$

where $D\mathscr{G}_{\tau(q,r,s)+(1-\tau)(\tilde{q},\tilde{r},\tilde{s})}$ is the Jacobian matrix at any point of the segment joining (q, r, s) and $(\tilde{q}, \tilde{r}, \tilde{s})$, $|\cdot|_2$ is the Euclidean norm in $\mathbb{R}^3$, and $\|\cdot\|_2$ is the operator norm induced by the previous norm $|\cdot|_2$, i.e., the norm such that for any arbitrary three-by-three matrix M is defined by

$$\|M\|_2 = \sup\{|M(q, r, s)'|_2 : (q, r, s) \in \mathbb{R}^3 \quad \text{such that} \quad |(q, r, s)|_2 = 1\}.$$

For this particular case, the matrix norm $\|\cdot\|_2$ is the same as the so-called spectral matrix norm [18] and $M(q, r, s)'$ denotes a matrix of finite dimension. The spectral norm of a matrix M is the square root of the largest eigenvalue of the positive-semidefinite matrix M^*M, i.e.

$$\|M\|_2 = \sqrt{\lambda_{max}(M^*M)},$$

where M^* is the conjugate transpose of M. Let first obtain the Jacobian matrix of $\mathscr{G}$,

$$D\mathscr{G}_{(q,r,s)} = \begin{pmatrix} bke^{-ms}\cos(kq) & 0 & -bme^{-ms}\sin(kq) \\ -dke^{-ms}\sin(kq) & 0 & -dme^{-ms}\cos(kq) \\ -ake^{-ms}\sin(kq) & 0 & -ame^{-ms}\cos(kq) \end{pmatrix}.$$

Now, for this real matrix, $D\mathscr{G}^*_{(q,r,s)}D\mathscr{G}_{(q,r,s)}$ is given by

$$e^{-2ms} \times$$

$$\begin{pmatrix} b^2k^2\cos^2(kq) + (a^2+d^2)k^2\sin^2(kq) & 0 & (a^2+d^2-b^2)km\sin(kq)\cos(kq) \\ 0 & 0 & 0 \\ (a^2+d^2-b^2)km\sin(kq)\cos(kq) & 0 & b^2m^2\sin^2(kq) + (a^2+d^2)m^2\cos^2(kq) \end{pmatrix}$$

$$= e^{-2ms}\begin{pmatrix} b^2k^2 & 0 & 0 \\ 0 & 0 & 0 \\ 0 & 0 & b^2m^2 \end{pmatrix} = b^2 e^{-2ms}\begin{pmatrix} k^2 & 0 & 0 \\ 0 & 0 & 0 \\ 0 & 0 & m^2 \end{pmatrix},$$

where we take into account (2.3). Hence, $\{0,\ b^2k^2e^{-2ms},\ b^2m^2e^{-2ms}\}$ is the set of eigenvalues of $D\mathscr{G}^*_{(q,r,s)}D\mathscr{G}_{(q,r,s)}$. The maximum of this set is readily obtained from the inequality (2.5). Consequently,

$$\|D\mathscr{G}_{(q,r,s)}\|_2 = \sqrt{b^2m^2e^{-2ms}} = bme^{-ms}.$$

On the other hand,

$$\max_{\tau\in[0,1]} \|D\mathscr{G}_{\tau(q,r,s)+(1-\tau)(\tilde{q},\tilde{r},\tilde{s})}\|_2 = bme^{-m\bar{s}},$$

where $\bar{s}$ is the s-component of the point on the line segment joining (q,r,s) and $(\tilde{q},\tilde{r},\tilde{s})$ that maximises bme^{-ms}. In particular, $\bar{s} \geq s_0$. Now by (3.6),

$$|\mathscr{G}(q,r,s) - \mathscr{G}(\tilde{q},\tilde{r},\tilde{s})|_2 \leq bme^{-m\bar{s}}|(q,r,s) - (\tilde{q},\tilde{r},\tilde{s}))|_2$$

$$\leq bme^{-ms_0}|(q,r,s) - (\tilde{q},\tilde{r},\tilde{s}))|_2.$$

Returning to the function $\mathscr{F}$, the previous inequality yields

$$|\mathscr{F}(q,r,s) - \mathscr{F}(\tilde{q},\tilde{r},\tilde{s})|_2 = |(q,r,s) - (\tilde{q},\tilde{r},\tilde{s}) - \mathscr{G}(q,r,s) + \mathscr{G}(\tilde{q},\tilde{r},\tilde{s})|_2$$

$$= |(q,r,s) - (\tilde{q},\tilde{r},\tilde{s}) - (\mathscr{G}(q,r,s) - \mathscr{G}(\tilde{q},\tilde{r},\tilde{s}))|_2$$

$$\geq |(q,r,s) - (\tilde{q},\tilde{r},\tilde{s})|_2 - |\mathscr{G}(q,r,s)) - \mathscr{G}(\tilde{q},\tilde{r},\tilde{s})|_2$$

$$\geq |(q,r,s) - (\tilde{q},\tilde{r},\tilde{s})|_2 - bme^{-ms_0}|(q,r,s) - (\tilde{q},\tilde{r},\tilde{s})|_2$$

$$= (1 - bme^{-ms_0})|(q,r,s) - (\tilde{q},\tilde{r},\tilde{s})|_2.$$

Therefore, when (3.5) holds, the following bound is obtained

$$|(q,r,s) - (\tilde{q},\tilde{r},\tilde{s})|_2 \leq \frac{1}{1 - bme^{-ms_0}}|\mathscr{F}(q,r,s) - \mathscr{F}(\tilde{q},\tilde{r},\tilde{s})|_2.$$

Finally, if two arbitrary points have the same image, then the norm of the difference of those two points is zero, and therefore, the two points must be the same. Thus, $\mathscr{F}$ is globally injective on $\overline{\mathscr{D}}$. □

Now, $\mathscr{F}$ is shown to be a homeomorphism by means of the Invariance of Domain Theorem 3.1, whose proof within Degree Theory can be found in [33] or in a more elementary way in [25].

Theorem 3.1 (Invariance of Domain) *If $U \subset \mathbb{R}^n$ is an open set and $f : \overline{U} \to \mathbb{R}^n$ is a continuous one-to-one mapping, then $f(U)$ is open.*

In the previous theorem, the result is true for any open set and so the map is open. Finally, as a consequence, we have that $f : \overline{U} \to f(\overline{U})$ is a homeomorphism. Taking into account this we proceed to the main result of this paper.

Theorem 3.2 *If* $1 - bme^{-ms_0} > 0$ *then the map $\mathscr{F}$ is a global diffeomorphism from $\mathscr{D}$ into the interior of fluid region $\mathcal{M}(t)$ and $\mathscr{F}$ maps $\partial\mathscr{D}$ onto the boundary of $\mathcal{M}(t)$.*

Proof By Proposition 3.2, the function $\mathscr{F}$ is (globally) injective if

$$1 - bme^{-ms_0} > 0. \tag{3.7}$$

Furthermore, it is easy to check that $\mathscr{F}$ is continuous on $\overline{\mathscr{D}}$. Hence, by the invariance of domain theorem, the map

$$\mathscr{F} : \overline{\mathscr{D}} \to \mathscr{F}(\overline{\mathscr{D}})$$

is a homeomorphism. In particular, $\mathscr{F}(\partial\mathscr{D}) = \partial\mathscr{F}(\mathscr{D})$ and $\overline{\mathscr{F}(\mathscr{D})} = \mathscr{F}(\overline{\mathscr{D}})$. Now, $\overline{\mathscr{F}(\mathscr{D})}$ is precisely the fluid region $\mathcal{M}(t)$. Thus, $\mathscr{F}$ maps the labelling domain surjectively into the fluid domain (see Fig. 4). Note here that it is possible to make use of the almost periodicity of $\mathscr{F}$ and show explicitly that $\partial\mathscr{F}(\mathscr{D}_\lambda) = \mathscr{F}(\partial\mathscr{D}_\lambda)$ where $\mathscr{D}_\lambda$ is as in (3.3). However, the Invariance of Domain theorem avoids the tedious calculations. On the other hand, as $b \geq a$, the condition (3.7) implies

$$1 - ame^{-ms_0} > 0,$$

which is precisely the condition in Proposition 3.1. It follows that $\mathscr{F}$ is a bijective local diffeomorphism from the open set $\mathscr{D}$ into the open set $\mathscr{F}(\mathscr{D})$; therefore, $\mathscr{F}$ is a global diffeomorphism.   □

Finally, in order to analyse the conditions imposed on the parameters, we obtain the following necessary condition for the injectivity of $\mathscr{F}$.

Lemma 3.1 *If $\mathscr{F}$ is injective, then the inequality*

$$bke^{-ms_0} \leq 1, \tag{3.8}$$

must hold.

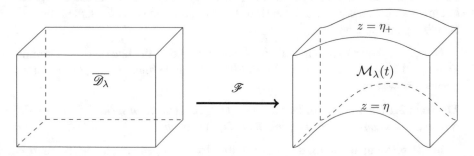

Fig. 4 Depiction of the transformation from the restricted labelling domain $\overline{\mathscr{D}_\lambda} = [0, 2\pi/k] \times [-r_0, r_0] \times [s_0(r), s_+(r)]$ to the correspondent fluid domain section of $\mathcal{M}_\lambda(t) = [0, 2\pi/k] \times [-y_0, y_0] \times [\eta, \eta_+]$

Proof Let us focus on the first component of (3.1). We define the following function of q alone

$$\mathcal{H}(q) := q - be^{-ms}\sin(kq),$$

and we look for solutions of the nonlinear equation given by

$$\mathcal{H}(q) = 0. \tag{3.9}$$

The proof of the Lemma is done by a contrapositive argument; therefore, let us assume that (3.8) does not hold, i.e.

$$bke^{-ms_0} > 1. \tag{3.10}$$

We have that $\mathcal{H}'(q) = 0$ if and only if,

$$q = \frac{1}{k}\arccos\left(\frac{1}{bke^{-ms_0}}\right).$$

The arccos in the previous equation is well-defined as long as (3.10) holds, and therefore, there exists a unique

$$q_0 \in \left(0, \frac{\pi}{2k}\right),$$

such that $\mathcal{H}'(q_0) = 0$. From the second derivative of the function $\mathcal{H}$, it follows that

$$\mathcal{H}''(q_0) = k\sqrt{b^2k^2e^{-2ms} - 1},$$

which is strictly positive by the same condition (3.10). Hence, there exists a unique critical point $q_0 \in (0, \frac{\pi}{2k})$ which is a minimum. Furthermore, $\mathcal{H}(0) = 0$ and $\mathcal{H}' < 0$ in a neighbourhood of zero; thus, $\mathcal{H}$ takes negatives values to the right of zero.

Finally,

$$\mathcal{H}(q) \longrightarrow \infty \quad \text{as} \quad q \to \infty. \tag{3.11}$$

To sum up, we have shown that

$$\mathcal{H}(0) = 0,$$

$$\mathcal{H}(q_0) < 0, \qquad \text{where } q_0 \text{ is a minimum for } \mathcal{H}(q) \text{ in } \left(0, \frac{\pi}{2k}\right),$$

which together with (3.11) shows that there exists a strictly positive solution of (3.9). If that solution is denoted by α, then it readily follows that $-\alpha$ is also a solution of

Fig. 5 Sketch of the graph of $\mathscr{H}$. Here q_0 is a local minimum for $\mathscr{H}$ and $\mathscr{H}(\alpha) = \mathscr{H}(-\alpha) = 0$

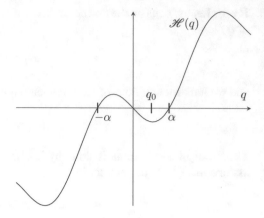

(3.9). This is schematically shown in Fig. 5. Turning our attention to the function $\mathscr{F}$, we show that it is not injective, as the points

$$(q_1, r_1, s_1) = (\alpha, r, s),$$

$$(q_2, r_2, s_2) = (-\alpha, r, s),$$

are such that

$$\mathscr{F}(q_1, r_1, s_1) = \left(0, \, r - de^{-ms} \cos(k\alpha), \, s - ae^{-ms} \cos(k\alpha)\right) = \mathscr{F}(q_2, r_2, s_2).$$

Therefore if $\mathscr{F}$ is injective then $bke^{-ms_0} \leq 1$. $\qquad\qquad\qquad\qquad\qquad\square$

Remark It is interesting to compare the different conditions imposed to $\mathscr{F}$. It has already been mentioned that $b \geq a$, thus the sufficient condition for the injectivity of $\mathscr{F}$ implies that $\mathscr{F}$ is an orientation-preserving local diffeomorphism. On the other hand, from Lemma 3.1 it follows that if $1 - bke^{-ms_0} < 0$, then $\mathscr{F}$ is not injective; however, from (2.4) $\mathscr{F}$ is a local diffeomorphism by the same type of argument used in Proposition 3.1 (although now the change of variables induced by $\mathscr{F}$ would have a strictly negative Jacobian and so it does not preserve the orientation).

In addition, along the equator we have that $a = b$ and $m = k$. Therefore, in this case, $\mathscr{F}$ is globally injective and a local diffeomorphism if and only if

$$1 - ame^{-ms_0} > 0.$$

Acknowledgements The authors acknowledge the support of the Erwin Schrödinger International Institute for Mathematics and Physics (ESI) during the program "Mathematical Aspects of Physical Oceanography" and the support of the Science Foundation Ireland (SFI) research grant 13/CDA/2117.

References

1. A. Bennett, *Lagrangian Fluid Dynamics* (Cambridge University Press, Cambridge, 2006)
2. J.P. Boyd, *Dynamics of the Equatorial Ocean* (Springer, Berlin, 2018)
3. A. Constantin, On the deep water wave motion. J. Phys. A: Math. Gen. **34**, 1405–1417 (2001)
4. A. Constantin, Edge waves along a sloping bed. J. Phys. A: Math. Gen **34**, 9723–9731 (2001)
5. A. Constantin, *Nonlinear Water Waves with Applications to Wave-Current Interactions and Tsunamis*, vol. 81 (SIAM, Philadelphia, 2011)
6. A. Constantin, An exact solution for Equatorially trapped waves. J. Geophys. Res. **117**, 1–8 (2012)
7. A. Constantin, Some three-dimensional nonlinear equatorial flows. J. Phys. Oceanogr. **43**, 165–175 (2013)
8. A. Constantin, Some nonlinear, equatorially trapped, nonhydrostatic internal geophysical waves. J. Phys. Oceanogr. **44**, 781–789 (2014)
9. A. Constantin, S.G. Monismith, Gerstner waves in the presence of mean currents and rotation. J. Fluid. Mech. **820**, 511–528 (2017). https://doi.org/10.1017/jfm.2017.223
10. B. Cushman-Roisin, J.M. Beckers, *Introduction to Geophysical Fluid Dynamics: Physical and Numerical Aspects*, vol. 101 (Academic Press, London, 2011)
11. F. Gerstner, Theorie der Wellen samt einer daraus abgeleiteten Theorie der Deichprofile. Ann. Phys. **2**, 412–445 (1809)
12. W. Froude, On the rolling of ships. Trans. Inst. Naval Arch. **2**, 45–62 (1862)
13. D. Henry, On Gerstner's water wave. J. Nonlinear Math. Phys. **15**, 87–95 (2008)
14. D. Henry, An exact solution for Equatorial geophysical water waves with an underlying current. Eur. J. Mech. B Fluids **38**, 190–195 (2013)
15. D. Henry, Equatorially trapped nonlinear water waves in a β-plane approximation with centripetal forces. J. Fluid Mech. **804** (2016). https://doi.org/10.1017/jfm.2016.544
16. D. Henry, A modified equatorial β-plane approximation modelling nonlinear wave-current interactions. J. Differ. Equ. **263**, 2554–2566 (2017)
17. D. Henry, On three-dimensional Gerstner-like equatorial water waves. Philos. Trans. R. Soc. A. Math. Phys. Eng. Sci. **376**, 1–16 (2017)
18. R.A. Horn, C.R. Johnson. *Matrix Analysis* (Cambridge University Press, Cambridge, 1985), p. 346
19. H.-C Hsu, An exact solution for nonlinear internal Equatorial waves in the f-plane approximation. J. Math. Fluid Mech. **16**, 463–471 (2014)
20. D. Ionescu-Kruse, An exact solution for geophysical edge waves in the f-plane approximation. Nonlinear Anal. Real World Appl. **24**, 190–195 (2015)
21. D. Ionescu-Kruse, On the short-wavelength stabilities of some geophysical flows. Philos. Trans. R. Soc. A. Math. Phys. Eng. Sci. **376**, 1–21 (2017)
22. R.S. Johnson, Application of the ideas and techniques of classical fluid mechanics to some problems in physical oceanography. Philos. Trans. R. Soc. A. Math. Phys. Eng. Sci. **376**, 1–19 (2017)
23. M. Kluczek, Exact and explicit internal Equatorially-trapped water waves with underlying currents. J. Math. Fluid Mech. **10**, 305–314 (2016)
24. M. Kluczek, Exact Pollard-like internal water waves. J. Nonlinear Math. Phys. **26**, 133–146 (2019)
25. W. Kulpa, Poincare and domain invariance theorem. Acta Univ. Carolin. Math. Phys. **39**, 127–136 (1998)
26. S. Monismith, H. Nepf, E. Cowen, L. Thais, J. Magnaudet, Laboratory observations of mean flows under surface gravity waves. J. Fluid Mech. **573**, 131–147 (2007)
27. A.V. Matioc, Exact geophysical waves in stratified fluids. Appl. Anal. **92**, 2254–2261 (2013)
28. R.T. Pollard, Surface waves with rotation: an exact solution. J. Geophys. Res **75**, 5895–5898 (1970)

29. W.J.M Rankine, On the exact form of waves near the surface of deep water. Philos. Trans. R. Soc. Lond. **153**, 127–138 (1863)
30. F. Reech, Sur la théorie des ondes liquides périodiques. C.R. Acad. Sci. Paris **68**, 1099–1101 (1869)
31. A. Rodríguez-Sanjurjo, Global diffeomorphism of the Lagrangian flow-map for Equatorially trapped internal water waves. Nonlinear Anal. **149**, 156–164 (2017)
32. A. Rodríguez-Sanjurjo, Global diffeomorphism of the Lagrangian flow-map for Pollard-like solutions. Ann. Mat. Pura Appl. **197**, 1787–1797 (2018)
33. E.H. Rothe, *Introduction to Various Aspects of Degree Theory in Banach Spaces. Mathematical Surveys and Monographs*, vol. 23 (American Mathematical Society, Providence, 1986)
34. S. Sastre-Gomez, Global diffeomorphism of the Lagrangian flow-map defining equatorially trapped water waves. Nonlinear Anal. **125**, 725–731 (2015)
35. R. Stuhlmeier, Internal Gerstner waves: applications to dead water. Appl. Anal. **93**, 1451–1457 (2014)

The Unified Transform and the Water Wave Problem

A. S. Fokas and K. Kalimeris

Abstract The unified transform, also known as the Fokas method, was introduced in 1997 by one of the authors Fokas (Proc R Soc Lond A: Math Phys Eng Sci 453(1962):1411–1443, 1997) for the analysis of nonlinear initial-boundary value problems. Later, it was realised that this method also yields novel results for linear problems. In 2006, the classical water wave problem was studied via the Fokas method (Ablowitz et al., J Fluid Mech 562:313–343, 2006), yielding a novel non-local formulation. In this paper we review the unified transform, with particular emphasis on its application in water wave in two spacial dimensions with moving boundaries.

Keywords Unified transform · Non-local formulation · Water waves

Mathematics Subject Classification (2000) Primary 35Q35; Secondary 76M40

1 Introduction

1.1 The Unified Transform

After the solution of the initial value problem for integrable evolution PDEs in one and two space dimensions, like the KdV and KP equations respectively, the most important open problem associated with the analysis of nonlinear integrable

A. S. Fokas
Department of Applied Mathematics and Theoretical Physics, University of Cambridge, Cambridge, UK

Viterbi School of Engineering, University of Southern California, Los Angeles, CA, USA
e-mail: tf227@cam.ac.uk

K. Kalimeris (✉)
Department of Applied Mathematics and Theoretical Physics, University of Cambridge, Cambridge, UK
e-mail: kk364@cam.ac.uk

© Springer Nature Switzerland AG 2019
D. Henry et al. (eds.), *Nonlinear Water Waves*, Tutorials, Schools, and Workshops in the Mathematical Sciences, https://doi.org/10.1007/978-3-030-33536-6_3

equations became the solution of initial-boundary value problems. A novel approach for the analysis of this problem was introduced by one of the authors in 1997 [19]. The linear limit of this approach gave rise to a new method for solving linear evolution PDEs [16, 22, 48]. Later it was realized that this method yields new integral representations for the solution of linear elliptic PDEs in polygonal domains [21], which in the case of simple domains can be used to obtain the analytical solution of several problems which apparently cannot be solved by the standard methods [40, 50]. In this way, a completely new method in mathematical physics emerged which is called unified transform or Fokas method [18, 20, 53]. An important role in this method is played by the so-called global relation which is an algebraic equation formulated in the complex Fourier plane, and which relates all relevant boundary values.

Although the global relation is only one of the ingredients of the Fokas method, still it has had important analytical and numerical implications: first, it has led to novel analytical formulations of a variety of important physical problems from water waves [15, 25, 46, 54] to three-dimensional layer scattering [45]. Second, it has led to the development of new numerical techniques for the Laplace, modified Helmholtz, Helmholtz, biharmonic equations [3, 7–9, 14, 28, 37–39], as well as for elliptic PDEs with variable coefficients [6].

Finally, it should be emphasised that the Fokas method has a significant pedagogical advantage: both the numerical calculation of the analytical solutions obtained for linear evolution PDEs, as well as the implementation of the numerical techniques to the elliptic PDEs are straightforward, so that even undergraduate students can implement then using MATLAB.

1.2   Water Waves

The study of water waves has been at the forefront of applied mathematics and engineering for over 200 years. This problem impacts a variety of important practical problems including harbour design, shipping, and tsunami prediction [44]. The problem formulated directly from the governing equations derived from physical principles is prohibitively difficult, because it requires the solution of Laplace's equation in an unknown domain, which is itself determined by nonlinear boundary conditions which depend on the solution. The reformulation of this problem in terms of the global relation presented in [1] reduces the problem to the solution of a global relation and a Bernoulli-like equation, vastly reducing its complexity. This result has had a significant impact in this classical area. In particular, it has led to new computations of surface water waves, and the discovery of new instabilities of waves in shallow water [15]. In addition, the employment of the global relation has inspired the incorporation of large amplitude effects in the reconstruction of the surface wave profile using pressure data measured at the bottom [46], as well as the solution of the inverse water wave problem, namely the determination of the bottom topography from surface wave data [54]. Recently, the authors have extended the results of [1] to investigate tsunami generation, by considering the water-wave problem with moving bottom [25].

2 The Global Relation

2.1 Integrable Problems

There exist two well known approaches to the exact analysis of linear PDEs: first, separation of variables gives rise to ordinary differential operators; the spectral analysis of these operators yields an appropriate transform pair. However, for non self-adjoint problems such transforms generally do not exist. The prototypical such pair is the Fourier transform; variations include the sine, the cosine, the Laplace and the Mellin transforms. Second, the use of integral representations obtained via Green's functions.

In the second half of the twentieth century it was realised that certain *nonlinear evolution* PDEs, called *integrable*, can be formulated as the compatibility condition of two linear eigenvalue equations called a *Lax pair*, and that this formulation gives rise to a method for solving the initial value problem for these equations, called the *inverse scattering transform* method. One of the authors has emphasised that this method is based on a deeper form of separation of variables [23]. Indeed, the spectral analysis of the t-independent part of the Lax pair yields an appropriate *nonlinear Fourier transform pair*, whereas the t-dependent part of the Lax pair yields the time evolution of the nonlinear Fourier data. In this sense, in spite of the fact that the inverse scattering transform is applicable to nonlinear PDEs, this method still follows the logic of separation of variables.

The *unified transform*, is based on two novel ideas (steps): (1) *Perform the simultaneous spectral analysis of both equations defining the Lax pair of the given PDE -or equivalently of a certain closed 1-differential form*—(this is to be contrasted with the case of initial value problems, where the spectral analysis of only the t-independent part of the Lax pair is performed). (2) *Analyse the global relation which couples the given initial and boundary data with the unknown boundary values*. The unified transform goes *beyond* separation of variables. Indeed, since it is based on the *simultaneous* spectral analysis of both parts of the Lax pair, it corresponds to the *synthesis* as opposed to separation of variables. As a consequence of this fundamental difference, even in the case of linear PDEs the form of the solution obtained by the unified transform differs drastically from the classical representations. It should be noted that the integral representations obtained classically via Green's functions, retain global features. Actually, it is shown in [35] and [51] that in the case of linear PDEs an alternative way to construct the novel integral representations obtained by the unified transform is to use appropriate contour deformations and Cauchy's theorem starting from the integral representations obtained via Green's functions (instead of performing the simultaneous spectral analysis of the associate Lax pair). In this sense, the unified transform reveals a deep relationship between the seminal contributions of Fourier, Cauchy and Green and furthermore extends these contributions to integrable *nonlinear* PDEs. Indeed, it is shown in [34] that for linear PDEs this method provides a unification as well as a significant extension of the classical transforms, of the method of images, of the Green's functions representations, and of the Wiener–Hopf technique (the latter technique through a series of ingenious steps gives rise to a Wiener–Hopf factorization problem, which is actually equivalent to a Riemann–Hilbert problem; in the new method, such Riemann–Hilbert problems can be *immediately* obtained using the global relation). Furthermore, the new approach provides an appropriate "nonlinearisation" of some of the above concepts.

It is well known that the main difficulty with boundary value problems stems from the fact that, although the solution representation requires the knowledge of all boundary values, some of them are *not* prescribed as boundary conditions. In the theory of elliptic PDEs, the determination of the unknown boundary values is known as the problem of characterising the generalised Dirichlet to Neumann map. For certain simple domains, the unified transform yields analytical expressions for the unknown boundary values [2, 4, 5, 10–13, 21, 26, 27, 32, 33]. For more complicated domains, it is remarkable that the unified transform yields a novel numerical technique for the determination of the unknown boundary values. For elliptic PDEs formulated in the interior of a polygon, this technique provides the analogue of the so-called "*boundary integral method*", but now the analysis takes place in the Fourier instead of the physical plane.

The problem of characterising the Dirichlet to Neumann map also appears in the analysis of initial boundary value problems for evolution equations formulated in either a fixed or a moving boundary. In the former case, using the unified transform it is possible to eliminate directly the unknown boundary values (or more precisely appropriate transforms of the unknown boundary values), and hence the problem of determining the generalised Dirichlet to Neumann map is bypassed [22, 29–31, 41, 47, 49, 52]. This is also possible for particular types of boundary conditions for integrable nonlinear PDEs; these boundary conditions are called linearisable. For general boundary conditions, the Dirichlet to Neumann map for integrable requires the analysis of certain nonlinear integral equations. Recently, significant progress has been made in the large t analysis of these equations [36, 42, 43].

2.2   *The Laplace Equation*

Let ϕ satisfy the Laplace equation, namely

$$\phi_{xx} + \phi_{yy} = 0. \tag{2.1}$$

Let

$$z = x + iy, \qquad \bar{z} = x - iy.$$

Using the equations

$$\frac{\partial}{\partial z} = \frac{1}{2}\left(\frac{\partial}{\partial x} - i\frac{\partial}{\partial y}\right), \qquad \frac{\partial}{\partial \bar{z}} = \frac{1}{2}\left(\frac{\partial}{\partial x} + i\frac{\partial}{\partial y}\right), \tag{2.2}$$

the Laplace equation (2.1) can be rewritten in the form

$$\phi_{z\bar{z}} = 0. \tag{2.3}$$

This equation immediately implies

$$\left(e^{-i\lambda z}\phi_z\right)_{\bar{z}} = 0, \quad \lambda \in \mathbb{C}, \tag{2.4}$$

which states that the function $e^{-i\lambda z}\phi_z$, is an analytic function. Hence, Cauchy's theorem yields the global relation

$$\int_{\partial\Omega} e^{-i\lambda z}\phi_z dz = 0, \quad \lambda \in \mathbb{C}. \tag{2.5}$$

Let Ω be the interior of the polygonal domain specified by the complex numbers z_1, $z_2, \ldots, z_n, z_{n+1} = z_1$ (Fig. 1).

Let L_j denote the side (z_j, z_{j+1}).

Let ϕ_T and ϕ_N denote the values of the directional and normal derivatives on each side of the polygon. Using the first of Eq. (2.2) and the identities

$$\phi_x dx + \phi_y dy = \phi_T ds \tag{2.6}$$

Fig. 1 Part of a polygonal domain

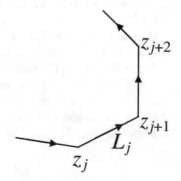

and

$$\phi_x dy - \phi_y dx = \phi_N ds \tag{2.7}$$

we can rewrite $\phi_z dz$ in a from involving only ϕ_T and ϕ_N:

$$\phi_z dz = \frac{1}{2}(\phi_x - i\phi_y)(dx + idy) = \frac{1}{2}(\phi_x dx + \phi_y dy) + \frac{1}{2}(\phi_x dy - \phi_y dx)$$

$$= \frac{1}{2}\phi_T ds + \frac{i}{2}\phi_N ds.$$

Hence, the global relation (2.5) becomes

$$\sum_{j=1}^{n} \hat{N}_j - i \sum_{j=1}^{n} \hat{D}_j = 0, \quad \lambda \in \mathbb{C}, \tag{2.8}$$

where $\{\hat{N}_j\}_1^n$ denote the transforms of the Neumann boundary values and $\{\hat{D}_j\}_1^n$ denote the transforms of the Dirichlet boundary values:

$$\hat{N}_j = \int_{z_j}^{z_{j+1}} e^{-i\lambda z}\phi_{N_j} ds, \quad j = 1, 2, \ldots, n, \quad \lambda \in \mathbb{C} \tag{2.9}$$

and

$$\hat{D}_j = \int_{z_j}^{z_{j+1}} e^{-i\lambda z}\phi_{T_j} ds, \quad j = 1, 2, \ldots, n, \quad \lambda \in \mathbb{C}. \tag{2.10}$$

Thus, the global relation involves n unknown functions, since for a well posed problem only one boundary condition is given on each side. This situation appears ominous, however in Eq. (2.8) the complex constant λ is *arbitrary*, thus Eq. (2.8) provides a family of equations. Using the symmetries of the global relation, analytical representations of the solutions are provided for simple domains such as the equilateral and the right isosceles triangle, see [40, 50].

Equation (2.4) which is a family of divergence forms, provides the linear analogue of the Lax pair appearing in the analysis of integrable equations.

It was shown in [17] that for a given elliptic PDE they exist several different global relations. For example, for the Laplace equation it also exists a global relation which involves the values of ϕ and ϕ_N on the boundary, instead of ϕ_T and ϕ_N. Thus, for a Dirichlet problem, where ϕ is given, the latter global relation is more convenient than (2.5).

3 Water Waves with Moving Boundaries

In this section we derive the equations for the free surface problem for two-dimensional irrotational flows in a domain where the rest of the boundaries are solid but moving.

Let $\Omega(t)$ denote this domain with moving boundaries, depicted in Fig. 2 and S denote the free surface.

Denoting the velocity of the flow by (u, v) the Euler equations for inviscid flow of an incompressible fluid are written as follows:

$$
\begin{aligned}
u_x + v_y &= 0, \\
u_t + uu_x + vu_y &= -P_x, \\
v_t + uv_x + vv_y &= -P_y - g, \quad \text{in } \Omega(t),
\end{aligned}
\tag{3.1}
$$

where P is the pressure and g is the gravitational constant.

Since the flow is irrotational the following equation is valid

$$
v_x - u_y = 0, \quad \text{in } \Omega(t).
\tag{3.2}
$$

Furthermore, we have the following boundary conditions on the free surface

$$
\begin{aligned}
P &= P_{atm} \text{ on } S, \\
v &= \eta_t + u\eta_x \text{ on } S.
\end{aligned}
\tag{3.3}
$$

The first of Eq. (3.3) is the dynamic boundary condition which states that the motion of the air is decoupled from the motion of the water. The second of Eq. (3.3) is

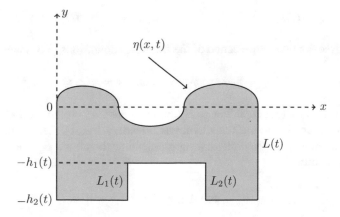

Fig. 2 Two dimensional water waves with piecewise horizontal moving bottom

the kinematic boundary condition which expresses the fact that the same particles always form the free water surface.

3.1 Mathematical Formulation of the Problem

Let ϕ denote the velocity potential, $(u, v) = \nabla \phi$.

The first of the Euler equations implies that the function $\phi(x, y, t)$ satisfies Laplace's equation in the domain $\Omega(t)$ (depicted by the grey area in Fig. 2) for $t > 0$, i.e.,

$$\phi_{xx} + \phi_{yy} = 0, \qquad (x, y) \in \Omega(t), \quad t > 0. \tag{3.4}$$

Furthermore let the left vertical be fixed, whereas all the other parts of the boundary are solid but moving. Then we obtain the following boundary conditions.

- On the vertical boundaries:

$$\phi_x(0, y, t) = 0, \qquad\qquad -h_2(t) < y < \eta(0, t), \quad t > 0, \tag{3.5}$$

$$\phi_x(L_1(t), y, t) = l_1(y, t), \qquad -h_2(t) < y < -h_1(t), \quad t > 0, \tag{3.6}$$

$$\phi_x(L_2(t), y, t) = r_1(y, t), \qquad -h_2(t) < y < -h_1(t), \quad t > 0, \tag{3.7}$$

$$\phi_x(L(t), y, t) = r(y, t), \qquad -h_2(t) < y < \eta(L(t), t), \quad t > 0. \tag{3.8}$$

- On the piecewise horizontal bottom:

$$\phi_y(x, -h_2(t), t) = b_1(x, t), \qquad 0 < x < L_1(t), \quad t > 0, \tag{3.9}$$

$$\phi_y(x, -h_1(t), t) = b(x, t), \qquad L_1(t) < x < L_2(t), \quad t > 0, \tag{3.10}$$

$$\phi_y(x, -h_2(t), t) = b_2(x, t), \qquad L_2(t) < x < L(t), \quad t > 0. \tag{3.11}$$

- On the free boundary the second of the boundary conditions (3.3) takes the form:

$$\eta_t + \eta_x \phi_x = \phi_y, \qquad \text{on} \quad y = \eta(x, t). \tag{3.12}$$

Equations (3.4)–(3.12) define a Neumann boundary value problem for the Laplace equation, involving the unknown boundary $\eta(x, t)$. The latter function can be determined, in principle, by supplementing the above equations with the additional condition

$$\phi_t + \frac{1}{2}(\phi_x^2 + \phi_y^2) + g\eta = 0, \text{ on } y = \eta(x, t). \tag{3.13}$$

This is the well-known Bernoulli's law, which is a consequence of the Euler equations. Indeed, condition (3.13) is derived by integrating the last equations of (3.1) and using Eq. (3.2); then we evaluate the resulting equation at the surface S and make usage of the first of the conditions (3.3).

We denote by $q(x, t)$ the value of $\phi(x, y, t)$, on the free surface, i.e.,

$$q(x, t) = \phi(x, \eta(x, t), t). \tag{3.14}$$

Our aim is to find the equations satisfied by $\eta(x, t)$ and $q(x, t)$. In this respect, differentiating (3.14) with respect to x, we obtain

$$q_x = \phi_x + \phi_y \eta_x, \qquad \text{on } y = \eta(x, t).$$

This equation together with Eq. (3.12) can be used to express the values of ϕ_x and ϕ_y at $y = \eta$, in terms of the following derivatives of q and η:

$$\phi_x = \frac{q_x - \eta_x \eta_t}{1 + \eta_x^2}, \quad \phi_y = \frac{\eta_t + \eta_x q_x}{1 + \eta_x^2}, \qquad \text{on } y = \eta(x, t). \tag{3.15}$$

Differentiating (3.14) with respect to t, we obtain

$$q_t = \phi_t + \phi_y \eta_t, \qquad \text{on } y = \eta(x, t).$$

Using in this equation the second of Eq. (3.15), we can express ϕ_t in terms of derivatives of q and η:

$$\phi_t = q_t - \frac{\eta_t + \eta_x q_x}{1 + \eta_x^2} \eta_t, \qquad \text{on } y = \eta(x, t).$$

Using the above expression together with Eq. (3.15) in condition (3.13), we obtain the first equation coupling q and η:

$$q_t + g\eta + \frac{1}{2} q_x^2 - \frac{1}{2} \frac{(\eta_t + \eta_x q_x)^2}{1 + \eta_x^2} = 0. \tag{3.16}$$

3.2 The Non-local Formulation

In order to obtain a second equation relating q and η, we introduce the complex variable $z = x + iy$. Then, following the analysis of Sect. 2.2 this equation comes as a result of the so-called global relation of the Laplace equation (2.3). Indeed, the global relation (2.5) provides a non-local formulation of the problem. In order to derive this formulation, we parametrise the boundary of the domain $\Omega(t)$.

On the free surface

$$z = x + i\eta(x, t).$$

Hence, using $dz = (1 + i\eta_x)dx$, and replacing ϕ_x, ϕ_y by Eq. (3.15), we find that the relevant contribution is given by the expression

$$\int_0^{L(t)} e^{i\lambda\left(x + i\eta(x,t)\right)} \left[\frac{q_x - \eta_x\eta_t}{1 + \eta_x^2} - i\frac{\eta_t + \eta_x q_x}{1 + \eta_x^2} \right] (1 + i\eta_x)dx.$$

This equation remarkably simplifies to the expression

$$\int_0^{L(t)} e^{i\lambda x - \lambda\eta} (q_x - i\eta_t) \, dx. \tag{3.17}$$

Using the boundary conditions (3.5)–(3.11) in Eq. (2.8), it is possible to compute the contribution of the remaining solid but moving boundary.

Thus, the global relation (2.5) takes the following form, see [25]

$$\int_0^{L(t)} e^{i\lambda x - \lambda\eta}(q_x - i\eta_t)dx - e^{\lambda h_1(t)} \int_{L_1(t)}^{L_2(t)} e^{i\lambda x} \phi_x(x, -h_1(t), t)dx$$

$$- e^{\lambda h_2(t)} \int_{L_2(t)}^{L(t)} e^{i\lambda x} \phi_x(x, -h_2(t), t)dx - e^{\lambda h_2(t)} \int_0^{L_1(t)} e^{i\lambda x} \phi_x(x, -h_2(t), t)dx \tag{3.18}$$

$$+ e^{i\lambda L_2(t)} \int_{-h_2(t)}^{-h_1(t)} e^{-\lambda y} \phi_y(L_2(t), y, t)dy - e^{i\lambda L_1(t)} \int_{-h_2(t)}^{-h_1(t)} e^{-\lambda y} \phi_y(L_1(t), y, t)dy$$

$$+ \int_{-h_2(t)}^{\eta(0,t)} e^{-\lambda y} \phi_y(0, y, t)dy - e^{i\lambda L(t)} \int_{-h_2(t)}^{\eta(L(t),t)} e^{-\lambda y} \phi_y(L(t), y, t)dy = iF(\lambda, t), \quad \lambda \in \mathbb{C},$$

where the known function $F(\lambda, t)$ is given by

$$F(\lambda, t) = e^{i\lambda L(t)} \int_{-h_2(t)}^{\eta(L(t),t)} e^{-\lambda y} r(y, t)dy - e^{\lambda h_1(t)} \int_{L_1(t)}^{L_2(t)} e^{i\lambda x} b(x, t)dx$$

$$- e^{\lambda h_2(t)} \int_0^{L_1(t)} e^{i\lambda x} b_1(x, t)dx - e^{\lambda h_2(t)} \int_{L_2(t)}^{L(t)} e^{i\lambda x} b_2(x, t)dx \tag{3.19}$$

$$+ e^{i\lambda L_1(t)} \int_{-h_2(t)}^{-h_1(t)} e^{-\lambda y} l_1(y, t)dy - e^{i\lambda L_2(t)} \int_{-h_2(t)}^{-h_1(t)} e^{-\lambda y} r_1(y, t)dy.$$

4 Water Waves with Moving Flat Bottom

In the case of a horizontal bottom, see Fig. 3, the global relation (3.18) simplifies. The non-local formulation is now obtained as a special case of the formulation (3.18)–(3.19), by making the substitutions

$$L_1(t) \equiv 0, \qquad L_2(t) \equiv L(t), \qquad h_2(t) \equiv h_1(t) =: h(t).$$

Thus, the boundary conditions (3.5)–(3.11) take the form

$$\phi_x(0, y, t) = 0, \qquad\qquad -h(t) < y < \eta(0, t), \quad t > 0, \qquad (4.1)$$

$$\phi_x(L(t), y, t) = r(y, t), \qquad -h(t) < y < \eta(L(t), t), \quad t > 0, \qquad (4.2)$$

$$\phi_y(x, -h(t), t) = b(x, t), \qquad 0 < x < L(t), \quad t > 0. \qquad (4.3)$$

Indeed, referring to [25] for details, the global relation (3.18) becomes

$$\int_0^{L(t)} e^{i\lambda x - \lambda \eta}(q_x - i\eta_t)dx - e^{\lambda h(t)} \int_0^{L(t)} e^{i\lambda x} \phi_x(x, -h(t), t)dx + \int_{-h(t)}^{\eta(0,t)} e^{-\lambda y} \phi_y(0, y, t)dy$$

$$- e^{i\lambda L(t)} \int_{-h(t)}^{\eta(L(t),t)} e^{-\lambda y} \phi_y(L(t), y, t)dy = iG(\lambda, t), \qquad \lambda \in \mathbb{C}, \qquad (4.4)$$

where the known function $G(\lambda, t)$ is given by

$$G(\lambda, t) = e^{i\lambda L(t)} \int_{-h(t)}^{\eta(L(t),t)} e^{-\lambda y} r(y, t)dy - e^{\lambda h(t)} \int_0^{L(t)} e^{i\lambda x} b(x, t)dx. \qquad (4.5)$$

It turns out that in the case of a horizontal bottom we can eliminate the unknown boundary values of the global relation (4.4), so that we can obtain a single equation coupling q and η, see Eq. (4.6) below.

Fig. 3 Two dimensional water waves with a horizontal (moving) bottom

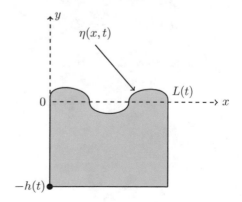

Indeed, it is shown in [25] that by using the invariances of the arbitrary parameter λ we obtain the following:

$$\int_0^{L(t)} \left\{ q_x \sin(\lambda x) \sinh \left[\lambda(\eta + h(t)) \right] \right.$$

$$\left. + \eta_t \cos(\lambda x) \cosh \left[\lambda(\eta + h(t)) \right] \right\} dx = S(\lambda, t), \quad \lambda = \frac{n\pi}{L(t)}, \quad (4.6)$$

where $S(\lambda, t)$ is defined by

$$S(\lambda, t) = - \cos(\lambda L(t)) \int_{-h(t)}^{\eta(L(t),t)} \cosh \left[\lambda(y + h(t)) \right] r(y, t) dy$$

$$+ \int_0^{L(t)} \cos(\lambda x) b(x, t) dx, \quad \lambda \in \mathbb{C}, \ t > 0. \quad (4.7)$$

4.1 Non-dimensional Variables

In order to study two interesting limits of Eqs. (3.16) and (3.18) we replace all variables with primed variables, and then we introduce dimensionless variables:

$$x' = L_0 x, \quad y' = h_0 y, \quad \eta' = \alpha \eta, \quad \lambda' = \frac{\lambda}{L_0}, \quad t' = \frac{L_0}{c_0} t, \quad q' = \epsilon c_0 L_0 q, \quad (4.8)$$

where α is a typical wave amplitude, and

$$\epsilon = \frac{\alpha}{h_0}, \quad \delta = \frac{h_0}{L_0}, \quad c_0^2 = g h_0. \quad (4.9)$$

Moreover, consistent asymptotics yield the following scaling

$$r'(y', t') = \frac{\alpha c_0}{\delta L_0} r(y, t), \qquad b'(x', t') = \frac{\alpha c_0}{L_0} b(x, t). \quad (4.10)$$

Using the above dimensionless variables, Eq. (3.16) becomes,

$$q_t + \eta + \frac{1}{2} \epsilon q_x^2 - \frac{1}{2} \epsilon \delta^2 \frac{(\eta_t + \epsilon \eta_x q_x)^2}{1 + \epsilon^2 \delta^2 \eta_x^2} = 0. \quad (4.11)$$

Equation (4.6) becomes

$$
\int_0^{\frac{L(t)}{L_0}} \left\{ q_x \sin(\lambda x) \frac{1}{\delta} \sinh\left[\lambda\delta\left(\frac{h(t)}{h_0}+\epsilon\eta\right)\right] + \eta_t \cos(\lambda x)\cosh\left[\lambda\delta\left(\frac{h(t)}{h_0}+\epsilon\eta\right)\right] \right\} dx
$$

$$
= \int_0^{\frac{L(t)}{L_0}} \cos(\lambda x)b(x,t)dx - (-1)^n \int_{-\frac{h(t)}{h_0}}^{\epsilon\eta\left(\frac{L(t)}{L_0},t\right)} \cosh\left[\lambda\delta\left(y+\frac{h(t)}{h_0}\right)\right] r(y,t)dy,
$$

$$
\lambda = \frac{n\pi L_0}{L(t)}.
$$

(4.12)

4.2 Boussinesq Type Equations

We note that small ϵ is indicative of small amplitude and small δ is indicative of long waves.

Next we fix the vertical boundary, namely

$$
L(t) = L_0, \qquad \text{and} \qquad r(y,t) = 0.
$$

In what follows we present two Boussinesq type equations: The first, which has a non-local form, is given in (4.13) and corresponds to the "small amplitude" approximation. The second one, which corresponds to the "small amplitude and long wave" approximation, is given in (4.17).

Letting $\epsilon \to 0$ in Eq. (4.12) and neglecting terms of $\mathcal{O}\left(\epsilon^2\right)$, we find a generalization of the Boussinesq equations, for a moving seafloor boundary, and, importantly, without the common "long wave" approximation:

$$
\eta_t + \epsilon \, (\eta q_x)_x + \sum_{n=0}^{\infty} \cos(\lambda x) \left\{ \tanh\left(\lambda\delta\frac{h}{h_0}\right) \int_0^1 \left[q_\xi \frac{\sin(\lambda\xi)}{\delta} + \epsilon\delta\lambda\eta\eta_t \cos(\lambda\xi) \right] d\xi \right\}
$$

$$
= \sum_{n=0}^{\infty} \frac{\cos(\lambda x)}{\cosh(\lambda\delta\frac{h}{h_0})} \int_0^1 b(\xi,t)\cos(\lambda\xi)d\xi, \qquad \text{for } \lambda = n\pi, \qquad (4.13)
$$

equipped with the boundary conditions

$$
q_x(0,t) = q_x(1,t) = 0.
$$

(4.14)

Furthermore, Eq. (4.11) becomes

$$
q_t + \eta + \frac{1}{2}\epsilon q_x^2 - \frac{1}{2}\epsilon\delta^2\eta_t^2 = 0.
$$

(4.15)

Under the additional assumption that δ is small, so that terms of $\mathcal{O}(\epsilon\delta^2)$ can be neglected, Eq. (4.11) becomes

$$q_t + \eta + \frac{\epsilon}{2}q_x^2 = 0, \tag{4.16}$$

and Eq. (4.12) yield the following well posed boundary value problem for the newly derived Boussinesq-type equation:

$$\eta_t + \frac{h(t)}{h_0}q_{xx} + \epsilon(\eta q_x)_x - \frac{\delta^2}{2}\left(\frac{h(t)}{h_0}\right)^2 \eta_{txx} - \frac{\delta^2}{6}\left(\frac{h(t)}{h_0}\right)^3 q_{xxxx} = -\frac{\dot{h}(t)}{h_0}, \tag{4.17}$$

along with the boundary conditions

$$q_x(0, t) = 0, \qquad q_x(1, t) = 0, \tag{4.18}$$

$$q_{xxx}(0, t) = 0, \qquad q_{xxx}(1, t) = 0. \tag{4.19}$$

The derivation of the above system of equations yields the conditions

$$\eta_{tx}(0, t) = \eta_{tx}(1, t) = 0.$$

4.3   Numerical Considerations

Considering small-amplitude water waves over a flat moving bottom, the employment of the scaling (4.8)–(4.10) in the global relation (4.4) yields the following equation, see [25]:

$$\int_0^{\frac{L(t)}{L_0}} e^{i\lambda x}\left[q_x - i\delta\eta_t - i\epsilon\delta(\eta q_x)_x - \epsilon\delta^2(\eta\eta_t)_x\right]dx - e^{\lambda\delta\frac{h(t)}{h_0}}\int_0^{\frac{L(t)}{L_0}} e^{i\lambda x}q_b(x, t)dx$$

$$+ \delta^2\int_{-\frac{h(t)}{h_0}}^0 e^{-\delta\lambda y}q_l(y, t)dy - \delta^2 e^{i\lambda\frac{L(t)}{L_0}}\int_{-\frac{h(t)}{h_0}}^0 e^{-\delta\lambda y}q_r(y, t)\,dy \tag{4.20}$$

$$= i\delta\left[-e^{\lambda\delta\frac{h(t)}{h_0}}\int_0^{\frac{L(t)}{L_0}} e^{i\lambda x}b(x, t)dx + e^{i\lambda\frac{L(t)}{L_0}}\int_{-\frac{h(t)}{h_0}}^0 e^{-\delta\lambda y}r(y, t)dy\right],$$

where we have neglected terms of $\mathcal{O}(\epsilon^2)$.

This equation is similar with the equation obtained for the analysis of the Laplace equation formulated in the interior of a rectangle with corners at

$$\left\{ 0, \ \frac{L(t)}{L_0}, \ \frac{L(t)}{L_0} - i\frac{h(t)}{h_0}, \ -i\frac{h(t)}{h_0} \right\},$$

but with two important differences: First, the boundary values on the side $\left(0, \frac{L(t)}{L_0}\right)$ are quadratically nonlinear, and second all boundary values depend on t. This implies that Eq. (4.20) can be integrated numerically via the method presented in [39], with the following modifications:

1. $q_x(x, t)$, $\eta(x, t)$, $q_b(x, t)$, $q_l(y, t)$, and $q_r(y, t)$ should be expanded in terms of Chebysev instead of Legendre polynomials, since Chebysev polynomials have the important property that

$$T_n(x)T_m(x) = \frac{1}{2}T_{n-m}(x) + \frac{1}{2}T_{n+m}(x).$$

2. The coefficients of the expansions are now functions of time.

Similar considerations are valid also for the case of the general case of piecewise horizontal moving bottom.

4.4 Water Waves with Fixed Boundaries

If we fix the flat bottom then the two Boussinesq-type equations (4.13) and (4.17) take the form

$$\eta_t + \epsilon (\eta q_x)_x \tag{4.21}$$

$$+ \sum_{n=0}^{\infty} \cos(\lambda x) \left\{ \tanh(\lambda\delta) \int_0^1 \left[q_\xi \frac{\sin(\lambda\xi)}{\delta} + \epsilon\delta\lambda\eta\eta_t \cos(\lambda\xi) \right] d\xi \right\} = 0, \qquad \lambda = n\pi.$$

and

$$\eta_t + q_{xx} + \epsilon(\eta q_x)_x - \frac{\delta^2}{2}\eta_{txx} - \frac{\delta^2}{6}q_{xxxx} = 0, \tag{4.22}$$

respectively.

We emphasise that the former equation corresponds to the "small amplitude" approximation, where the terms of $\mathcal{O}\left(\epsilon^2\right)$ are neglected, and the latter one corresponds to the "small amplitude and long wave" approximation, where the terms of $\mathcal{O}\left(\epsilon\delta^2\right)$ are neglected, too.

For a more detailed discussion of the limiting cases of the non-local formulation for water with fixed boundaries, in two and three spatial dimensions we refer to [24]. A hybrid of the novel formulation and an approach based on conformal mappings is presented in [28], where water waves with non-flat but fixed bottom are studied. We note that therein a Boussinesq-type equation is derived for the case of non-flat bottom, in the regime of "small amplitude and long wave"; the specific form of this equation for the flat bottom coincides with (4.22).

Acknowledgements The authors "A. S. Fokas and K. Kalimeris" were supported by EPSRC. The authors are grateful to the Erwin Schrödinger International Institute for Mathematics and Physics, Vienna, for the support and hospitality during the 2017 Nonlinear Water Waves—an Interdisciplinary Interface workshop.

References

1. M.J. Ablowitz, A.S. Fokas, Z.H. Musslimani, On a new non-local formulation of water waves. J. Fluid Mech. **562**, 313–343 (2006)
2. Y.A. Antipov, A.S. Fokas, March. The modified Helmholtz equation in a semi-strip, in *Mathematical Proceedings of the Cambridge Philosophical Society*, vol. 138(2) (Cambridge University Press, Cambridge, 2005), pp. 339–365
3. A.C.L. Ashton, The spectral Dirichlet–Neumann map for Laplace's equation in a convex polygon. SIAM J. Math. Anal. **45**(6), 3575–3591 (2013)
4. D. ben-Avraham, A.S. Fokas, The solution of the modified Helmholtz equation in a wedge and an application to diffusion-limited coalescence. Phys. Lett. A **263**(4–6), 355–359 (1999)
5. D. ben-Avraham, A.S. Fokas, Solution of the modified Helmholtz equation in a triangular domain and an application to diffusion-limited coalescence. Phys. Rev. E **64**(1), 016114 (2001)
6. M.J. Colbrook, Extending the unified transform: curvilinear polygons and variable coefficient PDEs. IMA J. Numer. Anal. (2018)
7. M.J. Colbrook, A.S. Fokas, Computing eigenvalues and eigenfunctions of the Laplacian for convex polygons. Appl. Numer. Math. **126**, 1–17 (2018)
8. M.J. Colbrook, N. Flyer, B. Fornberg, On the Fokas method for the solution of elliptic problems in both convex and non-convex polygonal domains. J. Comput. Phys. **374**, 996–1016 (2018)
9. M.J. Colbrook, A.S. Fokas, P. Hashemzadeh, A hybrid analytical-numerical technique for elliptic PDEs. SIAM J. Sci. Comput. **41**(2), A1066–A1090 (2019)
10. D.G. Crowdy, A.M. Davis, Stokes flow singularities in a two-dimensional channel: a novel transform approach with application to microswimming. Proc. R. Soc. A **469**(2157), 20130198 (2013)
11. D.G. Crowdy, A.S. Fokas, Explicit integral solutions for the plane elastostatic semi-strip, in *Proceedings of the Royal Society of London A: Mathematical, Physical and Engineering Sciences*, vol. 460(2045), (The Royal Society, London, 2004), pp. 1285–1309
12. G. Dassios, A.S. Fokas, The basic elliptic equations in an equilateral triangle, in *Proceedings of the Royal Society of London A: Mathematical, Physical and Engineering Sciences*, vol. 461(2061) (The Royal Society, London, 2005), pp. 2721–2748
13. G. Dassios, A.S. Fokas, Methods for solving elliptic PDEs in spherical coordinates. SIAM J. Appl. Math. **68**(4), 1080–1096 (2008)
14. C.I.R. Davis, B. Fornberg, A spectrally accurate numerical implementation of the Fokas transform method for Helmholtz-type PDEs. Complex Variables Elliptic Equ. **59**(4), 564–577 (2014)

15. B. Deconinck, K. Oliveras, The instability of periodic surface gravity waves. J. Fluid Mech. **675**, 141–167 (2011)
16. B. Deconinck, T. Trogdon, V. Vasan, The method of Fokas for solving linear partial differential equations. SIAM Rev. **56**(1), 159–186 (2014)
17. M. Dimakos, A.S. Fokas, The Poisson and the biharmonic equations in the interior of a convex polygon. Stud. Appl. Math. **134**(4), 456–498 (2015)
18. Fokas/Method. wikipedia.org/wiki/Fokas_method
19. A.S. Fokas, A unified transform method for solving linear and certain nonlinear PDEs, in *Proceedings of the Royal Society of London A: Mathematical, Physical and Engineering Sciences*, vol. 453(1962) (1997), pp.1411–1443
20. A.S. Fokas, On the integrability of linear and nonlinear partial differential equations. J. Math. Phys. 41(6), 4188–4237 (2000)
21. A.S. Fokas, Two-dimensional linear partial differential equations in a convex polygon, in *Proceedings of the Royal Society of London A: Mathematical, Physical and Engineering Sciences*, vol. 457(2006) (The Royal Society, London, 2001), pp. 371–393
22. A.S. Fokas, A new transform method for evolution partial differential equations. IMA J. Appl. Math. **67**(6), 559–590 (2002)
23. A.S. Fokas, Lax pairs: a novel type of separability (invited paper for the special volume of the 25th anniversary of Inverse Problems). Inverse Prob. **25**, 1–44 (2009)
24. A.S. Fokas, K. Kalimeris, *A Novel Non-Local Formulation of Water Waves*. Lectures on the Theory of Water Waves, vol. 426 (2016), p. 63
25. A.S. Fokas, K. Kalimeris, Water waves with moving boundaries. J. Fluid Mech. **832**, 641–665 (2017)
26. A.S. Fokas, A.A. Kapaev, A Riemann–Hilbert approach to the Laplace equation. J. Math. Anal. Appl. **251**(2), 770–804 (2000)
27. A.S. Fokas, A.A. Kapaev, On a transform method for the Laplace equation in a polygon. IMA J. Appl. Math. **68**(4), 355–408 (2003)
28. A.S. Fokas, A. Nachbin, Water waves over a variable bottom: a non-local formulation and conformal mappings. J. Fluid Mech. **695**, 288–309 (2012)
29. A.S. Fokas, D.T. Papageorgiou, Absolute and convective instability for evolution PDEs on the Half-line. Stud. Appl. Math. **114**(1), 95–114 (2005)
30. A.S. Fokas, B. Pelloni, Boundary value problems for Boussinesq type systems. Math. Phys. Anal. Geom. **8**(1), 59–96 (2005)
31. A.S. Fokas, B. Pelloni, A transform method for linear evolution PDEs on a finite interval. IMA J. Appl. Math. **70**(4), 564–587 (2005)
32. A.S. Fokas, D.A. Pinotsis, The Dbar formalism for certain linear non-homogeneous elliptic PDEs in two dimensions. Eur. J. Appl. Math. **17**(3), 323–346 (2006)
33. A.S. Fokas, D.A. Pinotsis, Quaternions, evaluation of integrals and boundary value problems. Comput. Methods Funct. Theory **7**(2), 443–476 (2007)
34. A.S. Fokas, E.A. Spence, Synthesis as opposed to separation of variables. SIAM Rev. **54**, 291–324 (2012)
35. A.S. Fokas, M. Zyskin, The fundamental differential form and boundary value problems. Q. J. Mech. Appl. Math. **55**, 457–479 (2002)
36. A.S. Fokas, J. Lenells, B. Pelloni, Boundary value problems for the elliptic sine-Gordon equation in a semi-strip. J. Nonlinear Sci. **23**(2), 241–282 (2013)
37. B. Fornberg, N. Flyer, A numerical implementation of Fokas boundary integral approach: Laplace's equation on a polygonal domain, in *Proceedings of the Royal Society of London A: Mathematical, Physical and Engineering Sciences* (The Royal Society, London, 2011), p. rspa20110032
38. E.N.G. Grylonakis, C.K. Filelis-Papadopoulos, G.A. Gravvanis, A class of unified transform techniques for solving linear elliptic PDEs in convex polygons. Appl. Numer. Math. **129**, 159–180 (2018)

39. P. Hashemzadeh, A.S. Fokas, S.A. Smitheman, A numerical technique for linear elliptic partial differential equations in polygonal domains, in *Proceedings Mathematical, Physical, and Engineering Sciences*, vol. 471(2175) (2015) p. 20140747
40. K. Kalimeris, Initial and boundary value problems in two and three dimensions, Doctoral Dissertation, University of Cambridge, 2010
41. K. Kalimeris, A.S. Fokas, The heat equation in the interior of an equilateral triangle. Stud. Appl. Math. **124**(3), 283–305 (2010)
42. J. Lenells, A.S. Fokas, The nonlinear Schrödinger equation with t-periodic data: I. Exact results. Proc. R. Soc. A **471**(2181), 20140925 (2015)
43. J. Lenells, J., A.S. Fokas, The nonlinear Schrödinger equation with t-periodic data: II. Perturbative results. Proc. R. Soc. A **471**(2181), 20140926 (2015)
44. S.P. Manual, *Coastal Engineering Research Center* (US Army Corps of Engineers, Washington, 1984)
45. D. Nicholls, A high-order perturbation of surfaces (HOPS) approach to Fokas integral equations: three-dimensional layered-media scattering. Q. Appl. Math. **74**(1), 61–87 (2016)
46. K. Oliveras, Stability of periodic surface gravity water waves. Doctoral dissertation, University of Washington, 2009
47. B. Pelloni, The spectral representation of two-point boundary-value problems for third-order linear evolution partial differential equations, in *Proceedings of the Royal Society of London A: Mathematical, Physical and Engineering Sciences*, vol. 461(2061) (The Royal Society, London, 2005), pp. 2965–2984
48. L.F. Rossi, Education. SIAM Rev. **56**(1), 157–158 (2014)
49. D.A. Smith, Well-posed two-point initial-boundary value problems with arbitrary boundary conditions, in *Mathematical Proceedings of the Cambridge Philosophical Society*, vol. 152(3) (Cambridge University Press, Cambridge, 2012), pp. 473–496
50. E.A. Spence, Boundary value problems for linear elliptic PDEs. Doctoral Dissertation, University of Cambridge, 2011
51. E. A. Spence, A.S. Fokas, A new transform method I: domain dependent fundamental solutions and integral representations. Proc. R. Soc. A. **466**, 2259–2281 (2010). A new transform method II: the global relation, and boundary value problems in polar co-ordinates. Proc. R. Soc. A. **466**, 2283–2307
52. T. Trogdon, B. Deconinck, The solution of linear constant-coefficient evolution PDEs with periodic boundary conditions. Appl. Anal. **91**(3), 529–544 (2012)
53. Unified Transform Method. unifiedmethod.azurewebsites.net
54. V. Vasan, B. Deconinck, The inverse water wave problem of bathymetry detection. J. Fluid Mech. **714**, 562–590 (2013)

HOS Simulations of Nonlinear Water Waves in Complex Media

Philippe Guyenne

Abstract We present an overview of recent extensions of the high-order spectral method of Craig and Sulem (J Comput Phys 108:73–83, 1993) to simulating nonlinear water waves in a complex environment. Under consideration are cases of wave propagation in the presence of fragmented sea ice, variable bathymetry and a vertically sheared current. Key components of this method, which apply to all three cases, include reduction of the full problem to a lower-dimensional system involving boundary variables alone, and a Taylor series representation of the Dirichlet–Neumann operator. This results in a very efficient and accurate numerical solver by using the fast Fourier transform. Two-dimensional simulations of unsteady wave phenomena are shown to illustrate the performance and versatility of this approach.

Keywords Bathymetry · Dirichlet–Neumann operator · Sea ice · Series expansion · Spectral method · Vorticity · Water waves

Mathematics Subject Classification (2000) Primary 76B15; Secondary 65M70

1 Introduction

The potential-flow formulation of Euler's equations for water waves has been very popular among both the mathematical and engineering communities, as it has proved to be successful at describing a wide range of wave phenomena. Via application of nonlocal operators, this formulation allows the original Laplace problem to be reduced from one posed inside the fluid domain to one posed on the boundary alone, thus allowing for dimensionality reduction. Moreover, in the absence of dissipative effects, the governing equations can be recast as a

P. Guyenne (✉)
Department of Mathematical Sciences, University of Delaware, Newark, DE, USA
e-mail: guyenne@udel.edu

© Springer Nature Switzerland AG 2019   53
D. Henry et al. (eds.), *Nonlinear Water Waves*, Tutorials, Schools, and Workshops
in the Mathematical Sciences, https://doi.org/10.1007/978-3-030-33536-6_4

canonical Hamiltonian system in terms of two conjugate variables, namely the surface elevation and the velocity potential evaluated there [40]. Due to these nice features, the potential-flow formulation has served as the theoretical basis in a countless number of water-wave studies, ranging from rigorous mathematical analysis to direct numerical simulation and weakly nonlinear modeling in various asymptotic regimes.

One of the most popular choices for direct numerical simulation is the so-called high-order spectral (HOS) approach, which is based on a Taylor series expansion of the Dirichlet–Neumann operator (DNO) combined with a pseudospectral scheme for space discretization using the fast Fourier transform. This is a very efficient and accurate numerical method when it is applicable. Compared to boundary integral methods [17, 19], it provides a faster recursive procedure for solving Laplace's equation in an irregular domain that is a perturbation to a simple geometry. Its computer implementation is also relatively easy and insensitive to the spatial dimension of the problem. From a general perspective, the basic idea underlying this approach is not marginal at all and, to some extent, shares similarities with other "fast" algorithms that are nowadays popular in scientific computing. For example, the fast multipole method [16] and more recently the method of quadrature by expansion [1] or the fast Chebyshev–Legendre transform [26] all rely on some sort of approximate series expansion in order to speed up computations. For the interested reader, details on boundary integral methods and other techniques can be found in other papers of this special volume.

The HOS approach was first introduced by Dommermuth and Yue [13] and West et al. [38] to simulate nonlinear gravity waves on uniform depth. Since then, it has been extended and applied to wave phenomena in various settings by many other investigators [14, 15, 28]. Slightly later than [13, 38], Craig and Sulem [6] proposed a related numerical method that has also been used with success in a number of subsequent applications [7, 8, 10, 30]. In particular, results were validated via comparison with laboratory experiments, weakly nonlinear predictions or other numerical solvers [9, 21–23, 39]. While these two HOS approaches are similar in their derivation, implementation and performance, there is a fundamental difference in their definition of the DNO. Dommermuth and Yue [13] and West et al. [38] define their DNO in terms of the vertical fluid velocity at the free surface, while Craig and Sulem [6] define their DNO in terms of the normal fluid velocity. These are two different quantities for a nontrivial free surface. In the latter definition, the DNO can be shown to be analytic with respect to surface deformations, which gives a justification for its Taylor series representation and thus a rigorous mathematical foundation for the corresponding HOS method [3]. Another important property of the DNO in that definition is its self-adjointness, which results in efficient and relatively simple recursion formulas for its computation [5, 30].

In an effort to improve the convergence of the DNO series, Nicholls and Reitich [32, 33] developed variants of Craig and Sulem's approach, which they refer to as Field Expansion and Transformed Field Expansion algorithms. These however require a hodograph transformation to map the irregular physical domain to a regular computational domain, together with a full-dimensional solution, because the elliptic problem becomes inhomogeneous. So far, they have only been used to compute traveling waves (i.e. steady waves in a moving reference frame) and to investigate the spectral stability of these solutions. A review on this body of work can be found in [31].

In this paper, we present an overview of recent work by the author and collaborators, that extends Craig and Sulem's approach to wave propagation in a complex environment. Most of these results have been obtained in the past decade or so, with a focus on unsteady solutions in the time domain. More specifically, we present direct numerical simulations of nonlinear dispersive waves in the presence of (i) fragmented sea ice [24], (ii) bottom topography [21] and (iii) a background shear current [18]. All three problems go beyond the classical setting of wave propagation in a homogeneous medium, and are of practical relevance to the fields of oceanography and coastal engineering. In particular, problem (i) has experienced renewed interest due to the rapid decline of summer ice extent that has occurred in the Arctic Ocean over recent years. Problem (iii) has also drawn much attention lately, especially from the mathematical community [4, 37], because it represents a refinement of the standard potential-flow formulation, allowing for rotational water waves. Therefore, we now find it timely to write a review paper on these recent advances, even more so considering that we are not aware of any previous review specifically on the HOS technique proposed by Craig and Sulem [6].

In all three cases, the numerical algorithm is based on the same original principle, and thus inherits the same qualities of accuracy and efficiency. In case (i), a mixed continuum-piecewise representation of flexural rigidity is adopted to specify an irregular array of ice floes on water. The main objective here is to emulate wave attenuation by scattering through an inhomogeneous ice field, as it may occur in the oceanic marginal ice zone. In contrast to linear predictions [36], slow or fast wave decay is observed depending on wave and ice parameters. In case (ii), the DNO exhibits an additional component that can be expanded in terms of bottom deformations. The inherent smoothing character of the DNO with respect to water depth is clearly revealed in this series expansion through the recurring presence of a smoothing Fourier multiplier. As a result, both smooth and non-smooth bottom profiles can be accommodated by this HOS method. In case (iii), wave propagation in the presence of constant nonzero vorticity is considered. This type of vorticity corresponds to a background shear current with a linear profile in the vertical direction. In addition to the DNO, another nonlocal operator (the Hilbert transform) is required in order to define a stream function at the free surface. A Taylor series expansion is also introduced for the fast computation of this operator. For an adverse current in deep water, it is confirmed that the Benjamin–Feir instability of Stokes waves may be significantly enhanced and may lead to the formation of large rogue waves [15].

The remainder of this paper is organized as follows. Sections 2 and 3 recall the basic governing equations in the potential-flow formulation for nonlinear water waves on uniform depth, as well as the corresponding Hamiltonian reduction and numerical discretization. While our HOS approach is extensible to three dimensions [5, 11, 25, 30, 39], we focus here on the two-dimensional case. Section 4 presents numerical results for wave propagation in the three different settings mentioned above (fragmented sea ice, variable bottom and shear current), with each setting discussed separately. In each case, we highlight the main points in the extension of the classical formulation.

2   Mathematical Formulation

2.1   Governing Equations

We consider the motion of a free surface on top of a two-dimensional ideal fluid of uniform depth h. In Cartesian coordinates, the x-axis is the direction of wave propagation and the y-axis points upward. The free surface is assumed to be the graph of a function $y = \eta(x, t)$. For potential flow, the velocity field is given by $\mathbf{u} = (u, v)^\top = \nabla\varphi$ where $\varphi(x, y, t)$ denotes the velocity potential. In terms of these variables, the initial boundary value problem for irrotational water waves associated with the fluid domain

$$S(\eta) = \{x \in \mathbb{R}, -h < y < \eta(x, t)\},$$

can be stated as

$$\Delta\varphi = 0, \quad \text{in} \quad S(\eta), \tag{2.1}$$

$$\eta_t - \varphi_y + \varphi_x \eta_x = 0, \quad \text{at} \quad y = \eta(x, t), \tag{2.2}$$

$$\varphi_t + \frac{1}{2}(\varphi_x^2 + \varphi_y^2) + g\eta + P = 0, \quad \text{at} \quad y = \eta(x, t), \tag{2.3}$$

$$\varphi_y = 0, \quad \text{at} \quad y = -h, \tag{2.4}$$

where g is the acceleration due to gravity and P represents normal stresses acting on the free surface (here $P = 0$ except for the sea-ice case where it is meant to model the bending force exerted by the floating ice sheet). Note that subscripts are used as shorthand notation for partial or variational derivatives (i.e. $\varphi_t = \partial_t \varphi$).

Following [6, 40], the dimensionality of the Laplace problem (2.1)–(2.4) can be reduced by introducing the trace of the velocity potential on the free surface, $\xi(x, t) = \varphi(x, \eta(x, t), t)$ together with the Dirichlet–Neumann operator (DNO)

$$G(\eta) : \xi \longmapsto (-\eta_x, 1)^\top \cdot \nabla\varphi\big|_{y=\eta},$$

which is the singular integral operator that takes Dirichlet data ξ at $y = \eta(x, t)$, solves Laplace's equation (2.1) subject to (2.4), and returns the corresponding

Neumann data (i.e. the normal velocity at the free surface). If $P = 0$, the resulting equations can be expressed as a canonical Hamiltonian system

$$\begin{pmatrix} \eta_t \\ \xi_t \end{pmatrix} = \begin{pmatrix} 0 & 1 \\ -1 & 0 \end{pmatrix} \begin{pmatrix} H_\eta \\ H_\xi \end{pmatrix},$$

for the conjugate variables η and ξ, whose Hamiltonian

$$H = \frac{1}{2} \int_{-\infty}^{\infty} \left[\xi G(\eta)\xi + g\eta^2 \right] dx,$$

corresponds to the total energy that is conserved over time. These equations more explicitly read

$$\eta_t = G(\eta)\xi, \tag{2.5}$$

$$\xi_t = -g\eta - \frac{1}{2(1 + \eta_x^2)} \left[\xi_x^2 - (G(\eta)\xi)^2 - 2\xi_x \eta_x G(\eta)\xi \right]. \tag{2.6}$$

2.2 Dirichlet–Neumann Operator

Equations (2.5) and (2.6) form a closed system for the two unknowns η and ξ. The question now is how to determine $G(\eta)\xi$ given η and ξ at any time, so that the right-hand sides of (2.5) and (2.6) can be evaluated. In two dimensions, it is known that G is an analytic function of η if $\eta \in \text{Lip}(\mathbb{R})$ [3]. Consequently, for surface perturbations around the quiescent state $\eta = 0$, the DNO can be written in terms of a convergent Taylor series expansion

$$G(\eta) = \sum_{j=0}^{\infty} G_j(\eta), \tag{2.7}$$

where the Taylor polynomials G_j are homogeneous of degree j in η and, as shown in [5, 6], they can be determined recursively: for even $j > 0$,

$$G_j = G_0 D^{j-1} \frac{\eta^j}{j!} D - \sum_{\ell=2,\, \text{even}}^{j} D^\ell \frac{\eta^\ell}{\ell!} G_{j-\ell} - \sum_{\ell=1,\, \text{odd}}^{j-1} G_0 D^{\ell-1} \frac{\eta^\ell}{\ell!} G_{j-\ell}, \tag{2.8}$$

and, for odd j,

$$G_j = D^j \frac{\eta^j}{j!} D - \sum_{\ell=2,\, \text{even}}^{j-1} D^\ell \frac{\eta^\ell}{\ell!} G_{j-\ell} - \sum_{\ell=1,\, \text{odd}}^{j} G_0 D^{\ell-1} \frac{\eta^\ell}{\ell!} G_{j-\ell}, \tag{2.9}$$

where $D = -i\partial_x$ and $G_0 = D\tanh(hD)$ are Fourier multiplier operators. In the infinite-depth limit ($h \to +\infty$), G_0 reduces to $|D|$ but otherwise Eqs. (2.8) and (2.9) remain unchanged. Using (2.7) together with (2.8) and (2.9) requires that η be a smooth single-valued function of x and thus overturning waves with a multivalued profile are not permitted. These formulas provide an efficient and accurate Laplace solver that lies at the heart of our HOS scheme as outlined below.

3 Numerical Methods

3.1 Space Discretization

Assuming periodic boundary conditions in the periodic cell $x \in [0, L_m)$, we use a pseudo-spectral method based on the fast Fourier transform (FFT). This is a suitable choice for computing the DNO since each term in (2.7) consists of concatenations of Fourier multipliers with powers of η. Accordingly, both functions η and ξ are expanded in truncated Fourier series

$$\begin{pmatrix} \eta \\ \xi \end{pmatrix} = \sum_{k=-k_m}^{k_m} \begin{pmatrix} \widehat{\eta}_k \\ \widehat{\xi}_k \end{pmatrix} e^{ikx}.$$

The spatial derivatives and Fourier multipliers are evaluated in the Fourier space, while the nonlinear products are calculated in the physical space on a regular grid of N collocation points. For example, if we wish to apply the zeroth-order operator G_0 to a function ξ in the physical space, we first transform ξ to the Fourier space, apply the diagonal operator $k\tanh(hk)$ to the Fourier coefficients $\widehat{\xi}_k$ and then transform back to the physical space.

In practice, the DNO series (2.7) is also truncated to a finite number of terms M but, by analyticity, a small number of terms (typically $M < 10 \ll N$) is sufficient to achieve highly accurate results [30, 32, 39]. Note that formulas (2.8) and (2.9) are slightly different from those originally given in [6] regarding the order of application of the various operators. As pointed out in [5], the DNO is self-adjoint and therefore the adjoint formulas (2.8) and (2.9) are equivalent to the original ones. This property however has important consequences on the DNO implementation and on the computational efficiency of the HOS approach. These adjoint formulas allow us to store and reuse the G_j's as vector operations on ξ, instead of having to recompute them at each order when applied to concatenations of Fourier multipliers and powers of η. This results in faster calculations and the computational cost for evaluating (2.7) is estimated to be $O(M^2 N \log N)$ operations via the FFT. Aliasing errors are removed by zero-padding in the Fourier space [30].

3.2 Time Integration

Time integration of (2.5) and (2.6) is performed in the Fourier space, which is advantageous for two main reasons. First, solving the time evolution problem amounts to solving an ODE system for the Fourier coefficients $\widehat{\eta}_k$ and $\widehat{\xi}_k$ rather than a PDE system for η and ξ. As mentioned above, the spatial derivatives are computed with spectral accuracy via the FFT. Second, the linear terms can be solved exactly by the integrating factor technique [6, 22, 39].

For this purpose, we separate the linear and nonlinear parts in (2.5) and (2.6). Setting $\mathbf{v} = (\eta, \xi)^\top$, these equations can be expressed as

$$\partial_t \mathbf{v} = \mathcal{L}\mathbf{v} + \mathcal{N}(\mathbf{v}), \tag{3.1}$$

where the linear part $\mathcal{L}\mathbf{v}$ is defined by

$$\mathcal{L}\mathbf{v} = \begin{pmatrix} 0 & G_0 \\ -g & 0 \end{pmatrix} \begin{pmatrix} \eta \\ \xi \end{pmatrix},$$

and the nonlinear part $\mathcal{N}(\mathbf{v})$ is given by

$$\mathcal{N}(\mathbf{v}) = \begin{pmatrix} [G(\eta) - G_0]\xi \\ -\frac{1}{2(1+\eta_x^2)}\left[\xi_x^2 - (G(\eta)\xi)^2 - 2\xi_x \eta_x G(\eta)\xi\right] \end{pmatrix}.$$

The change of variables $\widehat{\mathbf{v}}_k(t) = \Theta(t)\widehat{\mathbf{w}}_k(t)$ in the Fourier space reduces (3.1) to

$$\partial_t \widehat{\mathbf{w}}_k = \Theta(t)^{-1} \widehat{\mathcal{N}}_k\left[\Theta(t)\widehat{\mathbf{w}}_k\right],$$

via the integrating factor

$$\Theta(t) = \begin{pmatrix} \cos\left(t\sqrt{gG_0}\right) & \sqrt{\frac{G_0}{g}} \sin\left(t\sqrt{gG_0}\right) \\ -\sqrt{\frac{g}{G_0}} \sin\left(t\sqrt{gG_0}\right) & \cos\left(t\sqrt{gG_0}\right) \end{pmatrix},$$

for $k \neq 0$, and

$$\Theta(t) = \begin{pmatrix} 1 & 0 \\ -gt & 1 \end{pmatrix}.$$

for $k = 0$. The resulting system only contains nonlinear terms and is solved numerically in time using the fourth-order Runge–Kutta method with constant step Δt. After converting back to $\widehat{\mathbf{v}}_k$, this scheme reads

$$\widehat{\mathbf{v}}_k^{n+1} = \Theta(\Delta t)\widehat{\mathbf{v}}_k^n + \frac{\Delta t}{6}\Theta(\Delta t)\left(f_1 + 2f_2 + 2f_3 + f_4\right),$$

where

$$f_1 = \widehat{\mathcal{N}}_k \left(\widehat{\mathbf{v}}_k^n \right),$$

$$f_2 = \Theta \left(-\frac{\Delta t}{2} \right) \widehat{\mathcal{N}}_k \left[\Theta \left(\frac{\Delta t}{2} \right) \left(\widehat{\mathbf{v}}_k^n + \frac{\Delta t}{2} f_1 \right) \right],$$

$$f_3 = \Theta \left(-\frac{\Delta t}{2} \right) \widehat{\mathcal{N}}_k \left[\Theta \left(\frac{\Delta t}{2} \right) \left(\widehat{\mathbf{v}}_k^n + \frac{\Delta t}{2} f_2 \right) \right],$$

$$f_4 = \Theta(-\Delta t) \widehat{\mathcal{N}}_k \left[\Theta(\Delta t) \left(\widehat{\mathbf{v}}_k^n + \Delta t f_3 \right) \right],$$

for the solution at time $t_{n+1} = t_n + \Delta t$.

In cases of large-amplitude or highly deformed waves, filtering is needed in order to stabilize the numerical solution so that it can be computed over a sufficiently long time. Otherwise, spurious high-wavenumber instabilities tend to develop, eventually leading to computation breakdown, unless prohibitively small time steps are specified. This issue may be related to ill-conditioning of the DNO in its series form or may be promoted by the specific nonlinearity of the problem [32]. As a remedy, we apply a hyperviscosity-type filter of the form $\exp(-36|k/k_m|^{36})$ to the Fourier coefficients $\widehat{\eta}_k$ and $\widehat{\xi}_k$ at each time step. Such a filter has been commonly employed in direct numerical simulations of nonlinear fluid flows by spectral methods [27], and its form ensures that only energy levels at high wavenumbers are significantly affected. Therefore, if sufficiently fine resolution is specified, this filtering technique can help suppress spurious instabilities while preserving the overall solution. It also further contributes to removal of aliasing errors and thus blends well into the pseudo-spectral scheme.

4 Applications

In this section, we present applications of our HOS method to wave propagation in a complex environment. Extensions of the mathematical formulation described in Sect. 2 are briefly discussed, and simulations are shown to illustrate the capability and performance of the numerical model. Unless stated otherwise, Eqs. (2.5) and (2.6) are non-dimensionalized such that $g = 1$.

4.1 Fragmented Sea Ice

Floating sea ice is viewed as a thin elastic plate according to the special Cosserat theory of hyperelastic shells [29, 35]. This is modeled by an additional pressure term of the form $P = F\sigma/\rho$ on the right-hand side of (2.6), where

$$F = \frac{1}{2}\left(\frac{\eta_{xx}}{(1+\eta_x^2)^{3/2}}\right)^3 + \frac{1}{\sqrt{1+\eta_x^2}}\partial_x\left[\frac{1}{\sqrt{1+\eta_x^2}}\partial_x\left(\frac{\eta_{xx}}{(1+\eta_x^2)^{3/2}}\right)\right],$$

with ρ being the fluid density and σ the coefficient of ice rigidity [22, 23]. A spatial distribution of ice floes can be specified in the physical domain by allowing the coefficient of ice rigidity to be a variable function in space, namely $f(x)\sigma/\rho$, whose amplitude varies between 0 (open water) and σ/ρ (pack ice).

To generate a fragmented ice cover of total length L_c, we first prescribe a regular array of N_f identical floes whose individual length is L_f and which are evenly distributed over some distance L_c. Then, to make this arrangement look more irregular (and thus more realistic), each floe is shifted by an amount $\theta L_f/2$ relative to its initial center of gravity, where θ is a random number uniformly distributed between -1 and 1. At the edges of each floe, the continuous transition between the two phases is made steep but smooth enough to clearly distinguish the individual floes while complying with the continuum character of the underlying formulation. We use a tanh-like profile for this phase transition.

Focusing on the shallow-water regime, the present setup features a domain of length $L_m = 1200$, with the ice cover lying between $x = 100$ and $x = 1100$ (hence $L_c = 1000$). The objective is to quantify the attenuation of solitary waves propagating over this distance, for various floe configurations defined by $(N_f, L_f) = (77, 4), (77, 8), (13, 60), (13, 72)$ and corresponding to ice concentrations $C = N_f L_f/L_c = 0.31, 0.62, 0.78, 0.94$ respectively. The numerical parameters are set to $\Delta t = 0.002$, $N = 8192$ and $M = 6$.

Figure 1 shows snapshots of η as a solitary wave of initial amplitude $a_0/h = 0.3$ travels across the ice field. A single realization of each of the floe settings is considered. Two distinct mechanisms contributing to wave attenuation seem to coexist: multiple wave reflections from the ice floes (most apparent in the short-floe configurations), and pulse spreading due to the presence of ice itself (most apparent in the long-floe configurations). For the sparsest floe configuration $(N_f, L_f) = (77, 4)$, the solitary wave is seen to travel essentially unaffected aside from a slight decrease in amplitude. By contrast, for $(N_f, L_f) = (77, 8)$ which has a high level of ice concentration and ice fragmentation, the incident wave quickly decays through backward radiation and pulse spreading.

To further quantify the observed attenuation, Fig. 2 depicts the L^2 norm of η as a function of time for all four floe settings. Motivated by linear predictions [36], the least-squares exponential fit to each data set is also presented as a reference. While the exponential fit performs reasonably well for $(N_f, L_f) = (77, 4)$ when attenuation is weak, it provides a poorer approximation to the numerical data when attenuation is stronger. This is especially apparent in the case $(N_f, L_f) = (77, 8)$ where the data seem to converge to a nonzero limit rather than to zero as time goes on. This behavior may be attributed to the well-known stability of solitary waves

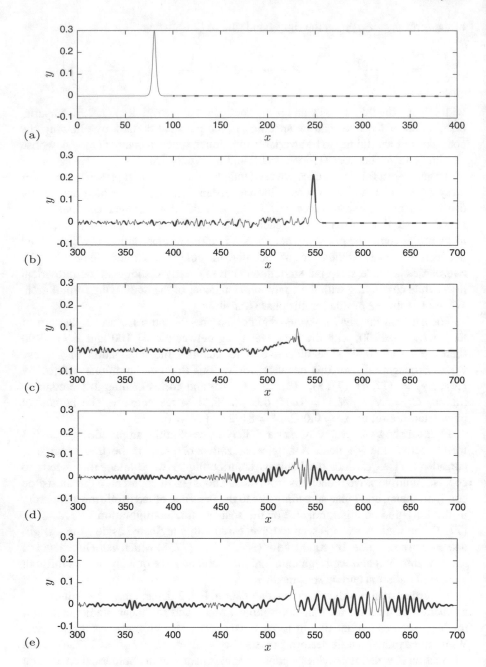

Fig. 1 Snapshots of η for $(N_f, L_f) = (77, 4)$ at $t = 0$ (**a**), $(77, 4)$ at $t = 416$ (**b**), $(77, 8)$ at $t = 416$ (**c**), $(13, 60)$ at $t = 416$ (**d**) and $(13, 72)$ at $t = 416$ (**e**) with $a_0/h = 0.3$. Open water is represented in blue while ice floes are represented in red

Fig. 2 L^2 norm of η as a function of time for $a_0/h = 0.3$. Numerical data are represented in various symbols while their exponential fits are plotted in solid line

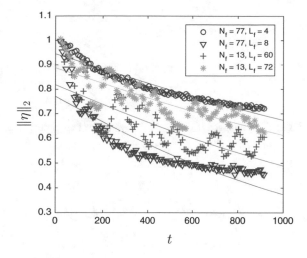

[9], which prevents them from completely disintegrating as they travel across the ice field. More details can be found in [24].

4.2 Bottom Topography

In this case, Eqs. (2.5) and (2.6) together with (2.7)–(2.9) can be used verbatim with the only exception that the first term G_0 is replaced by

$$G_0 = D \tanh(hD) + DL(\beta),$$

where $L(\beta)$ takes into account the bottom deformation $\beta(x)$ relative to a reference constant depth h [8, 20]. Because the DNO is jointly analytic with respect to β and η [34], $L(\beta)$ can be expressed in terms of a convergent Taylor series expansion in β,

$$L(\beta) = \sum_{j=0}^{\infty} \operatorname{sech}(hD) L_j(\beta), \tag{4.1}$$

where each L_j can be determined recursively: for even $j > 0$,

$$L_j = - \sum_{\ell=2,\,\text{even}}^{j-2} \frac{\beta^\ell}{\ell!} D^\ell L_{j-\ell} + \sum_{\ell=1,\,\text{odd}}^{j-1} \frac{\beta^\ell}{\ell!} \tanh(hD) D^\ell L_{j-\ell}, \tag{4.2}$$

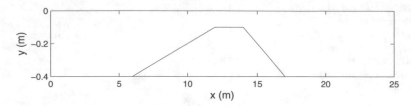

Fig. 3 Submerged bar in the Delft Hydraulics experiments [12]

and, for odd j,

$$L_j = -\frac{\beta^j}{j!}\operatorname{sech}(hD)D^j - \sum_{\ell=2,\,\text{even}}^{j-1}\frac{\beta^\ell}{\ell!}D^\ell L_{j-\ell} + \sum_{\ell=1,\,\text{odd}}^{j-2}\frac{\beta^\ell}{\ell!}\tanh(hD)D^\ell L_{j-\ell}.$$

(4.3)

These formulas clearly reveal the regularizing character of the DNO with respect to water depth, as indicated by the presence of the smoothing operator $\operatorname{sech}(hD)$. Any non-smoothness in the profile of β would automatically be regularized via action of the DNO, thus producing a C^∞ contribution [2, 10]. Adopting a Fourier series representation for β, Eqs. (4.1)–(4.3) are also evaluated by a pseudo-spectral method with the FFT. Similar to (2.7), the expansion (4.1) is truncated to a finite number of terms M_b that may be selected independently of M.

As an illustration, we consider the Delft Hydraulics bar experiments where a regular Stokes wave breaks up into higher harmonics after passing over a submerged bar [12]. As shown in Fig. 3, the bottom profile is not smooth and its amplitude is comparable to the total water depth. This case is particularly difficult to simulate because it involves wave propagation on deep and shallow water, over a wide range of depths. It has often been used as a discriminating test for nonlinear models of coastal waves. Figure 4 shows time series of η at various locations along the wave channel. At each location, our numerical results are compared with the experimental data. The incident wave has an amplitude $a_0 = 0.02\,\text{m}$ and period $T_0 = 2.02\,\text{s}$. The numerical parameters are set to $\Delta t = 0.001\,\text{s}$, $N = 2048$ and $M = M_b = 8$. Overall, the agreement between the two data sets is found to be quite good. In particular, the wave steepening during shoaling ($x < 13.5\,\text{m}$) and the generation of higher harmonics over the downslope of the bar ($x > 13.5\,\text{m}$) are well reproduced by the HOS model. More details can be found in [21], including the case of moving bottom topography.

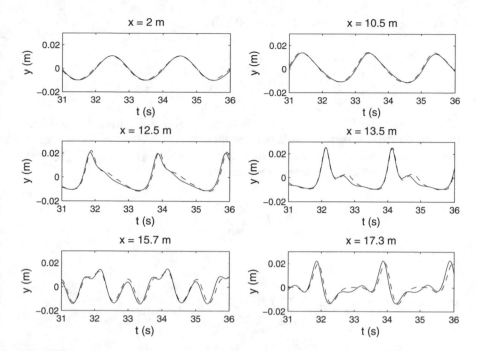

Fig. 4 Time series of η at various locations for an incident Stokes wave with $(a_0, T_0) = (0.02\,\text{m}, 2.02\,\text{s})$ passing over a bar: experiment (dashed line) and simulation (solid line)

4.3 Background Shear Current

In the presence of constant vorticity γ, the rotational flow can be described by two conjugate harmonic functions, namely a velocity potential φ and a stream function ψ, that satisfy

$$\varphi_x = \psi_y = u - U_0 + \gamma y, \qquad \varphi_y = -\psi_x = v,$$

where U_0 denotes a uniform background current [15, 37]. This leads to the following modifications in the Hamiltonian structure of the problem:

$$\begin{pmatrix} \eta_t \\ \xi_t \end{pmatrix} = \begin{pmatrix} 0 & 1 \\ -1 & \gamma \partial_x^{-1} \end{pmatrix} \begin{pmatrix} H_\eta \\ H_\xi \end{pmatrix},$$

with

$$H = \frac{1}{2} \int_{-\infty}^{\infty} \left[\xi G(\eta)\xi - \gamma \xi_x \eta^2 + \frac{1}{3}\gamma^2 \eta^3 - 2U_0 \xi \eta_x + g\eta^2 \right] dx.$$

The corresponding equations of motion take the form

$$\eta_t = G(\eta)\xi - U_0\eta_x + \gamma\eta\eta_x\,,$$

$$\xi_t = -g\eta - \frac{1}{2(1+\eta_x^2)}\left[\xi_x^2 - (G(\eta)\xi)^2 - 2\xi_x\eta_x G(\eta)\xi\right] - U_0\xi_x + \gamma\eta\xi_x - \gamma K(\eta)\xi\,,$$

where $K(\eta)\xi$, the Hilbert transform (HT) of ξ, returns the trace of the stream function on the free surface, i.e. $K(\eta)\xi = \psi(x, \eta(x, t), t)$. This is also a nonlocal operator that is related to the DNO by $G(\eta)\xi = -\partial_x K(\eta)\xi$. Similarly, it can be expressed in terms of a Taylor series expansion

$$K(\eta) = \sum_{j=0}^{\infty} K_j(\eta)\,,$$

where

$$K_j = -K_0 D^{j-2}\partial_x \frac{\eta^j}{j!}\partial_x + \sum_{\ell=2,\,\text{even}}^{j} D^{\ell-2}\partial_x \frac{\eta^\ell}{\ell!}\partial_x K_{j-\ell} + \sum_{\ell=1,\,\text{odd}}^{j-1} K_0 D^{\ell-1}\frac{\eta^\ell}{\ell!}\partial_x K_{j-\ell}\,,$$

for even $j > 0$, and

$$K_j = D^{j-1}\frac{\eta^j}{j!}\partial_x + \sum_{\ell=2,\,\text{even}}^{j-1} D^{\ell-2}\partial_x \frac{\eta^\ell}{\ell!}\partial_x K_{j-\ell} + \sum_{\ell=1,\,\text{odd}}^{j} K_0 D^{\ell-1}\frac{\eta^\ell}{\ell!}\partial_x K_{j-\ell}\,,$$

for odd j. The Fourier multiplier $K_0 = \mathrm{i}\tanh(hD)$ represents the HT for a uniform strip of thickness h. Because of this direct relation with the DNO, the same numerical procedure as described in Sect. 3.1 can be used to evaluate the HT series.

For simplicity, the following application only considers the case $U_0 = 0$. We investigate the Benjamin–Feir instability (BFI) of Stokes waves in the presence of a linear shear current. In the irrotational case ($\gamma = 0$), such waves are known to be unstable to sideband perturbations on deep water. We run simulations in a domain of length $L_m = 2\pi$ and infinite depth $h = +\infty$, with initial conditions representing a perturbed Stokes wave. The numerical parameters are set to $\Delta t = 0.001$, $N = 1024$ and $M = 6$. The initial Stokes wave has an amplitude $a_0 = 0.005$ with carrier wavenumber $k_0 = 10$, while the perturbation wavenumber is $\kappa = 1$.

Figure 5 shows snapshots of η at the initial time $t = 0$ for $\gamma = 0$ and at the time of maximum growth for $\gamma = 0, \pm 1, \pm 2$. We find that a co-propagating current ($\gamma > 0$) tends to stabilize the Stokes wave; the larger γ, the stronger the stabilizing effect. For $\gamma = +1$ and $+2$, the BFI seems to be inhibited and the corresponding graphs are not shown here because they look almost identical to Fig. 5a. On the other hand, a counter-propagating current ($\gamma < 0$) tends to promote and enhance the BFI. The larger $|\gamma|$, the sooner the Stokes wave becomes unstable and the higher it grows. For $\gamma = -1$ and -2, the wave reaches an elevation $a_{\max} = 0.016$ and $a_{\max} = 0.025$

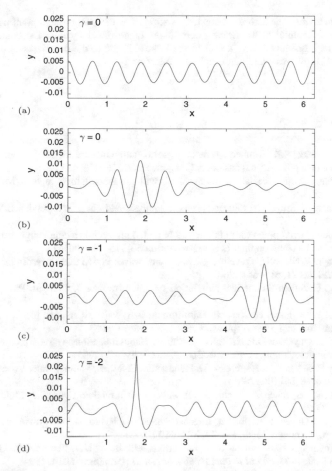

Fig. 5 Snapshots of η at (**a**) $t = 0$ ($\gamma = 0$), (**b**) $t = 956$ ($\gamma = 0$), (**c**) $t = 586$ ($\gamma = -1$) and (**d**) $t = 376$ ($\gamma = -2$) for an initially perturbed Stokes wave with $(a_0, k_0) = (0.005, 10)$ on deep water

at $t = 586$ and $t = 376$ respectively, which corresponds to an amplification factor of $\alpha = 3.2$ and $\alpha = 5$ compared to the initial amplitude a_0. As a reference, the maximum wave growth observed in Fig. 5b for $\gamma = 0$ is $\alpha = 2.4$ ($a_{\max} = 0.012$), which agrees with the classical NLS prediction

$$\alpha = \frac{a_{\max}}{a_0} = 1 + 2\sqrt{1 - \left(\frac{\kappa}{2\sqrt{2}k_0^2 a_0}\right)^2} = 2.4.$$

These results support the fact that wave-current interactions represent a possible mechanism for rogue wave formation in the ocean [15]. More details can be found in [18].

Acknowledgements The author acknowledges support by the NSF through grant number DMS-1615480. He is grateful to the Erwin Schrödinger International Institute for Mathematics and Physics for its hospitality during a visit in the fall 2017, and to the organizers of the workshop "Nonlinear Water Waves—an Interdisciplinary Interface".

References

1. L. af Klinteberg, A.K. Tornberg, A fast integral equation method for solid particles in viscous flow using quadrature by expansion. J. Comput. Phys. **326**, 420–445 (2016)
2. M. Cathala, Asymptotic shallow water models with non smooth topographies. Monatsh. Math. **179**, 325–353 (2016)
3. R. Coifman, Y. Meyer, Nonlinear harmonic analysis and analytic dependence. Proc. Symp. Pure Math. **43**, 71–78 (1985)
4. A. Compelli, Hamiltonian formulation of 2 bounded immiscible media with constant non-zero vorticities and a common interface. Wave Motion **54**, 115–124 (2015)
5. W. Craig, D.P. Nicholls, Traveling gravity water waves in two and three dimensions. Eur. J. Mech. B/Fluids **21**, 615–641 (2002)
6. W. Craig, C. Sulem, Numerical simulation of gravity waves. J. Comput. Phys. **108**, 73–83 (1993)
7. W. Craig, P. Guyenne, H. Kalisch, Hamiltonian long wave expansions for free surfaces and interfaces. Commun. Pure Appl. Math. **58**, 1587–1641 (2005)
8. W. Craig, P. Guyenne, D.P. Nicholls, C. Sulem, Hamiltonian long-wave expansions for water waves over a rough bottom. Proc. R. Soc. A **461**, 839–873 (2005)
9. W. Craig, P. Guyenne, J. Hammack, D. Henderson, C. Sulem, Solitary water wave interactions. Phys. Fluids **18**, 057106 (2006)
10. W. Craig, P. Guyenne, C. Sulem, Water waves over a random bottom. J. Fluid Mech. **640**, 79–107 (2009)
11. W. Craig, P. Guyenne, C. Sulem, Internal waves coupled to surface gravity waves in three dimensions. Commun. Math. Sci. **13**, 893–910 (2015)
12. M.W. Dingemans, Comparison of computations with Boussinesq-like models and laboratory measurements, in *Technical Report H1684.12* (Delft Hydraulics, Delft, 1994)
13. D.G. Dommermuth, D.K.P. Yue, A high-order spectral method for the study of nonlinear gravity waves. J. Fluid Mech. **184**, 267–288 (1987)
14. G. Ducrozet, F. Bonnefoy, D. Le Touzé, P. Ferrant, HOS-ocean: open-source solver for nonlinear waves in open ocean based on high-order spectral method. Comput. Phys. Commun. **203**, 245–254 (2016)
15. M. Francius, C. Kharif, S. Viroulet,Nonlinear simulations of surface waves in finite depth on a linear shear current, in *Proceedings of the 7th International Conference on Coastal Dynamics* (2013), pp. 649–660.
16. L. Greengard, V. Rokhlin, A fast algorithm for particle simulations. J. Comput. Phys. **73**, 325–348 (1987)
17. S.T. Grilli, P. Guyenne, F. Dias, A fully nonlinear model for three-dimensional overturning waves over an arbitrary bottom. Int. J. Numer. Meth. Fluids **35**, 829–867 (2001)
18. P. Guyenne, A high-order spectral method for nonlinear water waves in the presence of a linear shear current. Comput. Fluids **154**, 224–235 (2017)
19. P. Guyenne, S.T. Grilli, Numerical study of three-dimensional overturning waves in shallow water. J. Fluid Mech. **547**, 361–388 (2006)
20. P. Guyenne, D.P. Nicholls, Numerical simulation of solitary waves on plane slopes. Math. Comput. Simul. **69**, 269–281 (2005)
21. P. Guyenne, D.P. Nicholls, A high-order spectral method for nonlinear water waves over moving bottom topography. SIAM J. Sci. Comput. **30**, 81–101 (2007)

22. P. Guyenne, E.I. Părău, Computations of fully nonlinear hydroelastic solitary waves on deep water. J. Fluid Mech. **713**, 307–329 (2012)
23. P. Guyenne, E.I. Părău, Finite-depth effects on solitary waves in a floating ice sheet. J. Fluids Struct. **49**, 242–262 (2014)
24. P. Guyenne, E.I. Părău, Numerical study of solitary wave attenuation in a fragmented ice sheet. Phys. Rev. Fluids **2**, 034002 (2017)
25. P. Guyenne, D. Lannes, J.-C. Saut, Well-posedness of the Cauchy problem for models of large amplitude internal waves. Nonlinearity **23**, 237–275 (2010)
26. N. Hale, A. Townsend, A fast, simple, and stable Chebyshev–Legendre transform using an asymptotic formula. SIAM J. Sci. Comput. **36**, A148–A167 (2014)
27. T.Y. Hou, J.S. Lowengrub, M.J. Shelley, Removing the stiffness from interfacial flows with surface tension. J. Comput. Phys. **114**, 312–338 (1994)
28. Y. Liu, D.K.P. Yue, On generalized Bragg scattering of surface waves by bottom ripples. J. Fluid Mech. **356**, 297–326 (1998)
29. P.A. Milewski, Z. Wang, Three dimensional flexural-gravity waves. Stud. Appl. Math. **131**, 135–148 (2013)
30. D.P. Nicholls, Traveling water waves: spectral continuation methods with parallel implementation. J. Comput. Phys. **143**, 224–240 (1998)
31. D.P. Nicholls, Boundary perturbation methods for water waves. GAMM-Mitt. **30**, 44–74 (2007)
32. D.P. Nicholls, F. Reitich, Stability of high-order perturbative methods for the computation of Dirichlet–Neumann operators. J. Comput. Phys. **170**, 276–298 (2001)
33. D.P. Nicholls, F. Reitich, A new approach to analyticity of Dirichlet–Neumann operators. Proc. Roy. Soc. Edinburgh Sect. A **131**, 1411–1433 (2001)
34. D.P. Nicholls, M. Taber, Joint analyticity and analytic continuation of Dirichlet–Neumann operators on doubly perturbed domains. J. Math. Fluid Mech. **10**, 238–271 (2008)
35. P.I. Plotnikov, J.F. Toland, Modelling nonlinear hydroelastic waves. Phil. Trans. R. Soc. Lond. A **369**, 2942–2956 (2011)
36. P. Wadhams, V.A. Squire, D.J. Goodman, A.M. Cowan, S.C. Moore, The attenuation rates of ocean waves in the marginal ice zone. J. Geophys. Res. **93**, 6799–6818 (1988)
37. E. Wahlén, A Hamiltonian formulation of water waves with constant vorticity. Lett. Math. Phys. **79**, 303–315 (2007)
38. B.J. West, K.A. Brueckner, R.S. Janda, D.M. Milder, R.L. Milton, A new numerical method for surface hydrodynamics. J. Geophys. Res. **92**, 11803–11824 (1987)
39. L. Xu, P. Guyenne, Numerical simulation of three-dimensional nonlinear water waves. J. Comput. Phys. **228**, 8446–8466 (2009)
40. V.E. Zakharov, Stability of periodic waves of finite amplitude on the surface of a deep fluid. J. Appl. Mech. Tech. Phys. **9**, 190–194 (1968)

Stokes Waves
in a Constant Vorticity Flow

Sergey A. Dyachenko and Vera Mikyoung Hur

Abstract The Stokes wave problem in a constant vorticity flow is formulated via conformal mapping as a modified Babenko equation. The associated linearized operator is self-adjoint, whereby efficiently solved by the Newton-conjugate gradient method. For strong positive vorticity, a fold develops in the wave speed versus amplitude plane, and a gap as the vorticity strength increases, bounded by two touching waves, whose profile contacts with itself, enclosing a bubble of air. More folds and gaps follow as the vorticity strength increases further. Touching waves at the beginnings of the lowest gaps tend to the limiting Crapper wave as the vorticity strength increases indefinitely, while a fluid disk in rigid body rotation at the ends of the gaps. Touching waves at the boundaries of higher gaps contain more fluid disks.

Keywords Stokes wave · Constant vorticity · Conformal · Numerical

Mathematics Subject Classification (2000) Primary 76B15; Secondary 76B07, 30C30, 65T50

1 Introduction

Stokes in his classical treatise [20] (see also [21]) made formal but far-reaching considerations about periodic waves at the surface of an incompressible inviscid fluid in two dimensions, under the influence of gravity, which travel a long distance at a practically constant velocity without change of form. For instance, he observed that crests become sharper and troughs flatter as the amplitude increases, and that the 'wave of greatest height' exhibits a 120° corner at the crest. It would be impossible

S. A. Dyachenko
Department of Applied Mathematics, University of Washington, Seattle, WA, USA
e-mail: sdyachen@math.uiuc.edu

V. M. Hur (✉)
Department of Mathematics, University of Illinois at Urbana-Champaign, Urbana, IL, USA
e-mail: verahur@math.uiuc.edu

© Springer Nature Switzerland AG 2019
D. Henry et al. (eds.), *Nonlinear Water Waves*, Tutorials, Schools, and Workshops
in the Mathematical Sciences, https://doi.org/10.1007/978-3-030-33536-6_5

to give a complete account of Stokes waves here. We encourage the interested reader to some excellent surveys [2, 22, 24]. We merely pause to remark that in an irrotational flow of infinite depth, notable recent advances were based on a formulation of the problem as a nonlinear pseudodifferential equation, involving the periodic Hilbert transform, originally due to Babenko [1] (see also [10, 13, 17]). For instance, [3, 4] (see also [2] and references therein) rigorously addressed the existence in-the-large, and [11, 15, 16] numerically approximated the wave of greatest height and revealed the structure of complex singularities in great detail.

The irrotational flow assumption is well justified in some circumstances. But rotational effects are significant in many others, for instance, for wind driven waves, waves in a shear flow, or waves near a ship or pier. Constant vorticity is of particular interest because it greatly simplifies the mathematics. Moreover, for short waves, compared with the characteristic lengthscale of vorticity, the vorticity at the fluid surface would be dominant. For long waves, compared with the fluid depth, the mean vorticity would be dominant (see the discussion in [23]).

Simmen and Saffman [19] and Teles da Silva and Peregrine [23], among others, employed a boundary integral method and numerically computed Stokes waves in a constant vorticity flow. Their results include overhanging profiles and interior stagnation points. To compare, a Stokes wave in an irrotational flow is necessarily the graph of a single valued function and each fluid particle must move at a velocity less than the wave speed.

Recently, Constantin et al. [6] used conformal mapping, modified the Babenko equation and supplemented it with a scalar constraint, to permit constant vorticity and finite depth, and they rigorously established a global bifurcation result. The authors [8] rediscovered the modified Babenko equation and the scalar constraint, and numerically solved by means of the Newton-GMRES method (see also [5, 18]). More recently, the authors [9] eliminated the Bernoulli constant from the modified Babenko equation and, hence, the scalar constraint. The associated linearized operator is self-adjoint, whereby efficiently handled by means of the conjugate gradient method. Here we review the analytical formulation and numerical findings of [8, 9].

For strong positive vorticity, the amplitude increases, decreases and increases during the continuation of the numerical solution. Namely, a *fold* develops in the wave speed versus amplitude plane, and it becomes larger as the vorticity strength increases. For nonpositive vorticity, on the other hand, the amplitude increases monotonically. For stronger positive vorticity, a *gap* develops in the wave speed versus amplitude plane, bounded by two *touching waves*, whose profile contacts with itself at the trough line, enclosing a bubble of air, and the gap becomes larger as the vorticity strength increases. By the way, the numerical method of [19, 23] and others diverges in a gap. More folds and gaps follow as the vorticity strength increases even further.

Moreover, touching waves at the beginnings of the lowest gaps tend to the *limiting Crapper wave* (see [7]) as the vorticity strength increases indefinitely—a striking and surprising link between rotational and capillary effects—while they tend to a *fluid disk in rigid body rotation* at the ends of the gaps. Touching waves at the beginnings of the second gaps tend to the circular vortex wave on top of the limiting Crapper wave in the infinite vorticity limit, and the circular vortex wave on top of itself at the ends of the gaps. Touching waves at the boundaries of higher gaps contain more circular vortices in like manner.

2 Formulation

The water wave problem, in the simplest form, concerns the wave motion at the surface of an incompressible inviscid fluid in two dimensions, under the influence of gravity. Although an incompressible fluid may have variable density, we assume for simplicity that the density $= 1$. Suppose for definiteness that in Cartesian coordinates, the x axis points in the direction of wave propagation and the y axis vertically upward. Suppose that the fluid at time t occupies a region in the (x, y) plane, bounded above by a free surface $y = \eta(x, t)$ and below by the rigid bottom $y = -h$ for some constant h, possibly infinite. Let

$$\Omega(t) = \{(x, y) \in \mathbb{R}^2 : -h < y < \eta(x, t)\} \quad \text{and} \quad \Gamma(t) = \{(x, \eta(x, t)) : x \in \mathbb{R}\}.$$

Let $\boldsymbol{u} = \boldsymbol{u}(x, y, t)$ denote the velocity of the fluid at the point (x, y) and time t, and $P = P(x, y, t)$ the pressure. They satisfy the Euler equations for an incompressible fluid:

$$\boldsymbol{u}_t + (\boldsymbol{u} \cdot \nabla)\boldsymbol{u} = -\nabla P + (0, -g) \quad \text{and} \quad \nabla \cdot \boldsymbol{u} = 0 \quad \text{in } \Omega(t), \tag{1a}$$

where g is the constant due to gravitational acceleration. Let

$$\omega := \nabla \times \boldsymbol{u}$$

denote constant vorticity. By the way, if the vorticity is constant throughout the fluid at the initial time then Kelvin's circulation theorem implies that it remains so at later times. We assume that there is no motion in the air and we neglect the effects of surface tension. The kinematic and dynamic conditions:

$$\eta_t + \boldsymbol{u} \cdot \nabla(\eta - y) = 0 \quad \text{and} \quad P = P_{atm} \quad \text{at } \Gamma(t) \tag{1b}$$

express that each fluid particle at the surface remains so at all times, and that the pressure there equals the constant atmospheric pressure $= P_{atm}$. In the finite depth, $h < \infty$, the kinematic condition states

$$\boldsymbol{u} \cdot (0, -1) = 0 \quad \text{at } y = -h. \tag{1c}$$

We assume without loss of generality that the solutions of (1) are 2π periodic in the x variable.

For any $h \in (0, \infty)$, $\omega \in \mathbb{R}$ and $c \in \mathbb{R}$, clearly,

$$\eta(x, t) = 0, \quad u(x, y, t) = (-\omega y - c, 0) \quad \text{and} \quad P(x, y, t) = P_{atm} - gy \quad (2)$$

solve (1). We assume that some external effects such as wind produce such a constant vorticity flow and restrict the attention to waves propagating in (2).

Let

$$\boldsymbol{u} = (-\omega y - c, 0) + \nabla \Phi, \quad (3)$$

whence $\Delta \Phi = 0$ in $\Omega(t)$ by the latter equation of (1a). Naemly, Φ is a velocity potential for the irrotational perturbation from (2). For nonconstant vorticity, Φ is no longer viable to use. Let Ψ be a harmonic conjugate of Φ. Substituting (3) into the former equation of (1a), we make an explicit calculation to arrive at

$$\Phi_t + \frac{1}{2}(\Phi_x^2 + \Phi_y^2) - (\omega y + c)\Phi_x + \omega \Psi + P - P_{atm} + gy = b(t) \quad (4)$$

for some function $b(t)$. We substitute (3) into the other equations of (1), likewise. The result becomes, by abuse of notation,

$$\Delta \Phi = 0 \qquad\qquad\qquad\qquad\qquad\qquad \text{in } \Omega(t) \qquad (5a)$$

$$\eta_t + (\Phi_x - \omega \eta - c)\eta_x = \Phi_y \qquad\qquad \text{at } \Gamma(t), \qquad (5b)$$

$$\Phi_t + \frac{1}{2}|\nabla \Phi|^2 - (\omega \eta + c)\Phi_x + \omega \Psi + g\eta = 0 \qquad \text{at } \Gamma(t), \qquad (5c)$$

$$\Phi_y = 0 \qquad\qquad\qquad\qquad\qquad\qquad \text{at } y = -h. \qquad (5d)$$

By the way, since Φ and Ψ are determined up to arbitrary functions of t, we may take without loss of generality that $b(t) = 0$ at all times! In the infinite depth, $h = \infty$, we replace (5d) by

$$\Phi, \Psi \to 0 \quad \text{as } y \to -\infty \quad \text{uniformly for } x \in \mathbb{R}. \qquad (5e)$$

See [8, 9], for instance, for details.

2.1   Reformulations in Conformal Coordinates

To proceed, we reformulate (5) in conformal coordinates. Details may be found in [8, 9]. In what follows, we identify $\mathbb{R}^2$ with $\mathbb{C}$ whenever it is convenient to do so.

Let

$$z = z(w, t), \quad \text{where} \quad w = u + iv \quad \text{and} \quad z = x + iy, \qquad (6)$$

conformally map $\Sigma_d := \{u + iv \in \mathbb{C} : -d < v < 0\}$ of 2π period in the u variable, to $\Omega(t)$ of 2π period in the x variable, for some d, possibly infinite. Let (6) extend to map $\{u + i0 : u \in \mathbb{R}\}$ to $\Gamma(t)$, and $\{u - id : u \in \mathbb{R}\}$ to $\{x - ih : x \in \mathbb{R}\}$ if $d, h < \infty$, and $-i\infty$ to $-i\infty$ if $d, h = \infty$, where $d = \langle y \rangle + h$ (see [8] for detail). Here and elsewhere,

$$\langle f \rangle = \frac{1}{2\pi} \int_{-\pi}^{\pi} f(u) \, du$$

denotes the mean of a 2π periodic function f over one period.

Periodic Hilbert Transforms for a Strip For d in the range $(0, \infty)$, let

$$\mathcal{H}_d e^{iku} = -i \tanh(kd) e^{iku} \qquad\qquad \text{for } k \in \mathbb{Z}$$

and

$$\mathcal{T}_d e^{iku} = -i \coth(kd) e^{iku} \qquad\qquad \text{for } k \neq 0, \in \mathbb{Z}. \tag{7}$$

Let

$$\mathcal{H} e^{iku} = -i \operatorname{sgn}(k) e^{iku} \qquad\qquad \text{for } k \in \mathbb{Z}.$$

When $d < \infty$, if F is holomorphic in Σ_d and 2π periodic in the u variable and if $\operatorname{Re} F(\cdot + i0) = f$ and $(\operatorname{Re} F)_v(\cdot - id) = 0$ then

$$F(\cdot + i0) = (1 - i\mathcal{H}_d) f \tag{8}$$

up to the addition by a purely imaginary constant. Namely, $1 - i\mathcal{H}_d$ is the surface value of a periodic holomorphic function in a strip, the normal derivative of whose real part vanishes at the bottom. If $\operatorname{Im} F(\cdot + i0) = f$ and $\operatorname{Im} F(\cdot - id) = 0$, and if $\langle f \rangle = 0$, instead, then

$$F(\cdot + i0) = (\mathcal{T}_d + i) f \tag{9}$$

up to the addition by a real constant. Namely, $\mathcal{T}_d + i$ is the surface value of a periodic holomorphic function in a strip, whose imaginary part is of mean zero at the surface and vanishes at the bottom. Moreover, when $d = \infty$, if F is holomorphic in Σ_∞ and 2π periodic in the u variable and if F vanishes sufficiently rapidly at $-i\infty$ then the real and imaginary parts of $F(\cdot + i0)$ are the periodic Hilbert transforms for each other (see [26], for instance).

Implicit Form Note that $(x + iy)(u, t)$, $u \in \mathbb{R}$, makes a conformal parametrization of the fluid surface. In the finite depth, $d, h < \infty$, it follows from the Cauchy–

Riemann equations and (7) that

$$(x + iy)(u, t) = u + (\mathcal{T}_d + i)y(u, t). \tag{10}$$

In the infinite depth, $d, h = \infty$, $\mathcal{H}$ replaces $\mathcal{T}_d$ (see [9, 10], for instance).
    Moreover, let

$$(\phi + i\psi)(w, t) = (\Phi + i\Psi)(z(w, t), t) \quad \text{for } w \in \Sigma_d.$$

Namely, it is a conformal velocity potential for the irrotational perturbation from (2). In the finite or infinite depth, it follows from (8) that

$$(\phi + i\psi)(u, t) = (1 - i\mathcal{H}_d)\phi(u, t) \tag{11}$$

up to the addition by a purely imaginary constant.
    In the finite depth, substituting (10) and (11) into (5b) and (5c), we make an explicit calculation to arrive at

$$(1 + \mathcal{T}_d y_u)y_t - y_u \mathcal{T}_d y_t - \mathcal{H}_d \phi_u - (\omega y + c)y_u = 0,$$

$$\begin{aligned}((1 + \mathcal{T}_d y_u)^2 + y_u^2)(\phi_t + gy - \omega\mathcal{H}_d\phi) \\
- ((1 + \mathcal{T}_d y_u)\mathcal{T}_d y_t + y_u y_t)\phi_u + (y_u \mathcal{T}_d y_t - (1 + \mathcal{T}_d y_u)y_t)\mathcal{H}_d \phi_u \\
+ \frac{1}{2}(\phi_u^2 + (\mathcal{H}_d \phi_u)^2) - (\omega y + c)((1 + \mathcal{T}_d y_u)\phi_u - y_u \mathcal{H}_d \phi_u) = 0.\end{aligned} \tag{12}$$

In the infinite depth, $\mathcal{H}$ replaces $\mathcal{H}_d$ and $\mathcal{T}_d$. See [8], for instance, for details.

Explicit Form In the finite depth, note that z_t/z_u is holomorphic in Σ_d,

$$\text{Im}\frac{z_t}{z_u} = \frac{\mathcal{H}_d \phi_u + (\omega y + c)y_u}{|z_u|^2} \quad \text{at } v = 0$$

by the former equation of (12), and $\text{Im}(z_t/z_u) = 0$ at $v = -d$ by (5d). Note that $\langle \text{Im}(z_t/z_u) \rangle = 0$ for any $v \in [-d, 0]$ by the Cauchy–Riemann equations and (5d). It then follows from (9) that

$$\frac{z_t}{z_u} = (\mathcal{T}_d + i)\left(-\frac{(\mathcal{H}_d\phi + \frac{1}{2}\omega y^2 + cy)_u}{|z_u|^2}\right) \quad \text{at } v = 0. \tag{13}$$

Moreover, note that $(\phi_u - i\mathcal{H}_d\phi_u)^2$ is the surface value of a holomorphic and 2π periodic function in Σ_d, the normal derivative of whose real part vanishes at the bottom. It then follows from (8) and (7) that

$$\phi_u^2 - (\mathcal{H}_d\phi_u)^2 = -2\mathcal{T}_d(\phi_u \mathcal{H}_d\phi_u). \tag{14}$$

We use (13) and (14), and make a lengthy but explicit calculation to solve (12) as

$$y_t = (1 + \mathcal{T}_d y_u + y_u \mathcal{T}_d)\left(\frac{\mathcal{H}_d \phi_u + (\omega y + c) y_u}{(1 + \mathcal{T}_d y_u)^2 + y_u^2}\right),$$

$$\phi_t = -\phi_u \mathcal{T}_d\left(\frac{\mathcal{H}_d \phi_u + (\omega y + c) y_u}{(1 + \mathcal{T}_d y_u)^2 + y_u^2}\right)$$

$$+ \frac{1}{(1 + \mathcal{T}_d y_u)^2 + y_u^2}(\mathcal{T}_d(\phi_u \mathcal{H}_d \phi_u) + (\omega y + c)(1 + \mathcal{T}_d y_u)\phi_u) + \omega \mathcal{H}_d \phi - gy.$$

$$(15)$$

In the infinite depth, $\mathcal{H}$ replaces $\mathcal{H}_d$ and $\mathcal{T}_d$. See [8], for instance, for details.

2.2   The Stokes Wave Problem in a Constant Vorticity Flow

We turn the attention to the solutions of (15), for which y_t, $\phi_t = 0$.

In the finite depth, substituting $y_t = 0$ into the former equation of (15), we arrive at

$$\phi' = \mathcal{T}_d(\omega y y' + cy') \quad \text{at } v = 0. \qquad (16)$$

Here and elsewhere, the prime denotes ordinary differentiation. Substituting $\phi_t = 0$ into the latter equation of (15), likewise, we use (16) and we make an explicit calculation to arrive at

$$(c + \omega y(1 + \mathcal{T}_d y') - \omega \mathcal{T}_d(yy'))^2 = (c^2 - 2gy)((1 + \mathcal{T}_d y')^2 + (y')^2). \qquad (17)$$

In the infinite depth, $\mathcal{H}$ replaces $\mathcal{T}_d$. If we were to take (4), rather than (5c), where $b = 0$, then the result would become

$$(c + \omega y(1 + \mathcal{T}_d y') - \omega \mathcal{T}_d(yy'))^2 = (c^2 + 2b - 2gy)((1 + \mathcal{T}_d y')^2 + (y')^2), \qquad (18)$$

and one must determine b as part of the solution. See [8], for instance, for details.

The Modified Babenko Equation   Unfortunately, (17) or (18) is not suitable for numerical solution, because one would have to work with rational functions of y. We reformulate (17) as in a more convenient form. Details may be found in [8, 9].

In the finite depth, we rearrange (17) as

$$(c - \omega \mathcal{T}_d(yy'))^2 + 2\omega y(c - \omega \mathcal{T}_d(yy'))(1 + \mathcal{T}_d y') - \omega^2 y^2 (y')^2$$

$$= (c^2 - 2gy - \omega^2 y^2)((1 + \mathcal{T}_d y')^2 + (y')^2).$$

Note that $(c - \omega(\mathcal{T}_d + i)(yy'))^2$ is the surface value of a holomorphic and 2π periodic function in Σ_d, whose imaginary part is of mean zero at the surface and vanishes at the bottom. Hence, so is

$$(c^2 - 2gy - \omega^2 y^2)((1 + \mathcal{T}_d y')^2 + (y')^2) - 2\omega y(c - \omega \mathcal{T}_d(yy'))(1 + \mathcal{T}_d y' + iy')$$
$$= ((c^2 - 2gy - \omega^2 y^2)(1 + \mathcal{T}_d y' - iy') - 2\omega y(c - \omega \mathcal{T}_d(yy')))(1 + \mathcal{T}_d y' + iy').$$

Moreover, note that $1/(1 + \mathcal{T}_d y' + iy')$ is the surface value of the holomorphic and 2π periodic function $= 1/z_u$ in Σ_d, whose imaginary part is of mean zero at the surface and vanishes at the bottom. Hence, so is

$$(c^2 - 2gy - \omega^2 y^2)(1 + \mathcal{T}_d y' - iy') - 2\omega y(c - \omega \mathcal{T}_d(yy')).$$

Therefore, it follows from (10) that

$$(c^2 - 2gy - \omega^2 y^2)(1 + \mathcal{T}_d y') - 2\omega y(c - \omega \mathcal{T}_d(yy')) = -\mathcal{T}_d((c^2 - 2gy - \omega^2 y^2)y')$$

up to the addition by a real constant. Or, equivalently,

$$c^2 \mathcal{T}_d y' - (g + c\omega)y - g(y\mathcal{T}_d y' + \mathcal{T}_d(yy'))$$
$$- \frac{1}{2}\omega^2(y^2 + \mathcal{T}_d(y^2 y') + y^2 \mathcal{T}_d y' - 2y\mathcal{T}_d(yy')) = 0 \qquad (19)$$

and

$$g\langle y(1 + \mathcal{T}_d y')\rangle + c\omega\langle y\rangle + \frac{1}{2}\omega^2\langle y^2\rangle = 0. \qquad (20)$$

Indeed, $\langle \mathcal{T}_d f'\rangle = 0$ for any function f by (7) and

$$\langle y^2 \mathcal{T}_d y'\rangle = \frac{1}{2\pi}\int_{-\pi}^{\pi} y^2 \mathcal{T}_d y' \, du = -\frac{1}{2\pi}\int_{-\pi}^{\pi} y\mathcal{T}_d(y^2)' \, du = -\langle 2y\mathcal{T}_d(yy')\rangle.$$

In the infinite depth, $\mathcal{H}$ replaces $\mathcal{T}_d$. Conversely, a solution of (19) and (20) gives rise to a traveling wave of (5) and, hence, (1), provided that

$$u \mapsto (u + \mathcal{T}_d y(u), y(u)), u \in \mathbb{R}, \text{ is injective} \qquad (21a)$$

and

$$((1 + \mathcal{T}_d y_u)^2 + y_u^2)(u) \neq 0 \quad \text{for any } u \in \mathbb{R}. \qquad (21b)$$

See [8, 9], for instance, for details. The Stokes wave problem in a constant vorticity flow is to find $\omega \in \mathbb{R}, d \in (0, \infty], c \in \mathbb{R}$ an a 2π periodic function y, satisfying (21), which together solve (19) and (20). In what follows, we assume that y is even (see [12], for instance, for arbitrary vorticity).

In an irrotational flow of infinite depth, $\omega = 0$ and $d = \infty$, (19) and (20) simplify to

$$c^2 \mathcal{H} y' - gy - g(y\mathcal{H} y' + \mathcal{H}(yy')) = 0 \tag{22}$$

and $\langle y(1 + \mathcal{T}_d y') \rangle = 0$. Longuet-Higgins [13] discovered a set of identities among the Fourier coefficients of a Stokes wave, which Babenko [1] rediscovered in the form of (22) and, independently, [10, 17] among others. One may regard (19) and (20) as the modified Babenko equation, permitting constant vorticity and finite depth.

If we were to take (4), rather than (5c), where $b = 0$, then (19) would become

$$(c^2 + 2b)\mathcal{T}_d y' - (g + c\omega)y - g(y\mathcal{T}_d y' + \mathcal{T}_d(yy'))$$
$$- \frac{1}{2}\omega^2(y^2 + y^2\mathcal{T}_d y' + \mathcal{T}_d(y^2 y') - 2y\mathcal{T}_d(yy')) = 0, \tag{23}$$

which is supplemented with

$$\langle (c + \omega y(1 + \mathcal{T}_d y') - \omega \mathcal{T}_d(yy'))^2 \rangle = \langle (c^2 + 2b - 2gy)((1 + \mathcal{T}_d y')^2 + (y')^2) \rangle. \tag{24}$$

This is what [6, 8] derived.

3 Numerical Method

We write (19) in the operator form as $\mathcal{G}(y; c, \omega, d) = 0$ and solve it iteratively using the Newton method. Let $y^{(n+1)} = y^{(n)} + \delta y^{(n)}, n = 0, 1, 2, \ldots$, where $y^{(0)}$ is an initial guess, to be supplied (see [8, 9], for instance), and $\delta y^{(n)}$ solves

$$\delta \mathcal{G}(y^{(n)}; c, \omega, d)\delta y^{(n)} = -\mathcal{G}(y^{(n)}; c, \omega, d), \tag{25}$$

$\delta \mathcal{G}(y^{(n)}; c, \omega, d)$ is the linearization of $\mathcal{G}(y; c, \omega, d)$ with respect to y and evaluated at $y = y^{(n)}$.

We exploit an auxiliary conformal mapping, involving Jacobi elliptic functions (see [9] and references therein), and take efficient, albeit highly nonuniform, grid points in $u \in [-\pi, \pi]$. We approximate $y^{(n)}$ by a discrete Fourier transform and numerically evaluate $y^{(n)}$, $\mathcal{T}_d y^{(n)}$ and others using a fast Fourier transform. Since

$$\delta \mathcal{G}(y; c, \omega, d)\delta y = c^2 \mathcal{T}_d(\delta y)' - (g + c\omega)\delta y - g(\delta y \mathcal{T}_d y' + y \mathcal{T}_d(\delta y)' + \mathcal{T}_d(y\delta y)')$$
$$- \frac{1}{2}\omega^2 (2y\delta y + \mathcal{T}_d(y^2 \delta y)' - [2y\delta y, y] + [y^2, \delta y]),$$

where $[f_1, f_2] = f_1 \mathcal{T}_d f_2' - f_2 \mathcal{T}_d f_1'$, is self-adjoint, we solve (25) using the conjugate gradient (CG) method. We employ (20) to determine the zeroth Fourier coefficient. Once we arrive at a convergent solution, we continue it along in the parameters. See [9], for instance, for details.

If we were to take (23) and (24), rather than (19), where $b = 0$, then the associated linearized operator includes

$$(\delta y, \delta b) \mapsto (c^2 + 2b)\mathcal{T}_d(\delta y)' + 2\delta b \mathcal{T}_d y' - (g + c\omega)\delta y$$
$$- g(\delta y \mathcal{T}_d y' + y \mathcal{T}_d(\delta y)' + \mathcal{T}_d(y\delta y)')$$
$$- \frac{1}{2}\omega^2 (2y\delta y + \mathcal{T}_d(y^2 \delta y)' - [2y\delta y, y] + [y^2, \delta y]),$$

which is *not* self-adjoint, whence the CG or conjugate residual method may not apply. The authors [8] used the generalized minimal residual (GMRES) method and achieved some success. But it would take too much time to accurately resolve a numerical solution when it requires excessively many grid points. The CG method is more powerful than the GMRES for self-adjoint equations, and it leads to new findings, which we discuss promptly.

4   Results

Summarized below are the key findings of [8, 9].

We take without loss of generality that c is positive, and allow ω positive or negative, representing waves propagating upstream or downstream, respectively (see the discussion in [23]).

We take for simplicity that $g = 1$ and $d = \infty$. By the way, the effects of finite depth change the amplitude of a Stokes waves and others, but they are insignificant otherwise (see [8], for instance).

In what follows, the steepness s measures the crest-to-trough wave height divided by the period $= 2\pi$.

4.1   Folds and Gaps

For zero and negative constant vorticity, for instance, for $\omega = 0$ and -1, the left panel of Fig. 1 collects the wave speed versus steepness from the continuation of

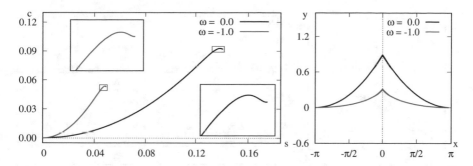

Fig. 1 On the left, wave speed vs. steepness for $\omega = 0$ and -1. Insets are closeups near the endpoints of the continuation of the numerical solution. On the right, the profiles of almost extreme waves

the numerical solution. For $\omega = 0$, Longuet-Higgins and Fox [14], among others, predicted that c oscillates infinitely many times whereas s increases monotonically toward the wave of greatest height or the extreme wave, whose profile exhibits a $120°$ corner at the crest. Numerical computations (see [11, 16], for instance, and references therein) bear it out. The insets reproduce the well-known result and suggest likewise when $\omega = -1$.

The right panel displays the profiles of almost extreme waves, in the (x, y) plane in the range $x \in [-\pi, \pi]$. Troughs are at $y = 0$. Note that the steepness when $\omega = -1$ is noticeably less than $\omega = 0$.

For a large value of positive constant vorticity, for instance, for $\omega = 2.5$, Fig. 2 includes the wave speed versus steepness and the profiles at the indicated points along the $c = c(s)$ curve, in the (x, y) plane, where $x \in [-3\pi, 3\pi]$. Troughs are at $y = 0$. The upper left panel reveals that s increases and decreases from $s = 0$ to wave D. Namely, a *fold* develops in the $c = c(s)$ curve. For s small, for instance, for wave A, the profile is single valued. But we observe that the profile becomes more rounded as s increases along the fold, so that overhanging waves appear, whose profile is no longer single valued. Moreover, we arrive at a *touching wave*, whose profile becomes vertical and contacts with itself somewhere the trough line, whereby enclosing a bubble of air. Wave B is an almost touching wave.

Past the touching wave, a numerical solution is unphysical because (21a) no longer holds true (see [8, 9] for examples). Moreover, we observe that the profile becomes less rounded as s decreases along the fold, so that we arrive at another touching wave; past the touching wave, a numerical solution is physical. Wave C is an almost touching wave and wave D is physical. Together, a *gap* develops in the $c = c(s)$ curve, consisting of unphysical numerical solutions and bounded by two touching waves. We remark that wave C encloses a larger bubble of air than wave B.

Past the end of the fold, interestingly, the upper left panel reveals another fold and another gap. The steepness increases from waves D to F, and decreases from waves F to H. Waves E and G are almost touching waves and numerical solutions

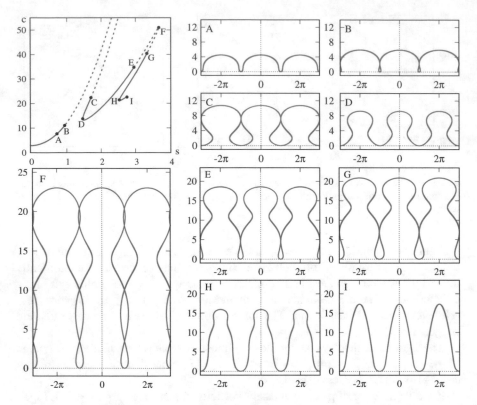

Fig. 2 For $\omega = 2.5$. Clockwise from upper left: wave speed vs. steepness; the profiles of eight solutions, labelled by A to E, and G to I; the profile of an unphysical solution labelled by F

between are unphysical. For instance, for wave F, the profile intersects itself and the fluid region overlaps itself.

Past the end of the second fold, we observe that s increases monotonically, although c oscillates (see [9], for instance, for details), like when $\omega = 0$; moreover, overhanging profiles disappear as s increases and the crests become sharper, like when $\omega = 0$. Therefore, we may claim that an *extreme wave* ultimately appears, whose profile exhibits a sharp corner at the crest. Wave I is an almost extreme wave. One may not continue the numerical solution past the extreme wave because (21b) would no longer hold true.

Figure 3 includes the wave speed versus steepness for several values of positive constant vorticity. For zero vorticity, one predicts that c experiences infinitely many oscillations whereas s increases monotonically (see [14], for instance). For negative constant vorticity, numerical computations (see [8, 19, 23], among others) suggest that the crests become sharper and lower. Figure 1 bears it out.

For positive constant vorticity, for instance, for $\omega = 1.7$, on the other hand, Fig. 3 reveals that the lowest oscillation of c deforms into a fold. Consequently, there correspond two or three solutions for some values of s. Moreover, the extreme

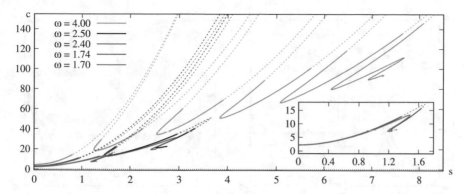

Fig. 3 Wave speed vs. steepness for five values of positive constant vorticity. Solid curves for physical solutions and dashed curves unphysical. The inset distinguishes the lowest fold and gap

wave seems not the wave of greatest height. We observe that the fold becomes larger in size as ω increases. For a larger value of the vorticity, for instance, for $\omega = 1.74$, the figure reveals that part of the fold transforms into a gap. We observe that the gap becomes larger in size as ω increases. See [8], for instance, for details.

Moreover, for $\omega = 2.4$, Fig. 3 reveals that the second oscillation of c deforms into another fold, and we observe that the second fold becomes larger in size as ω increases. For $\omega = 2.5$, part of the second fold transforms into another gap, and we observe that the second gap becomes larger in size as ω increases. We merely pause to remark that the numerical method of [19, 23] and others diverges in a gap and is incapable of locating a second gap. The numerical method of [8] converges in a gap, but it would take too much time to accurately resolve a numerical solution along a second fold.

We take matters further and claim that higher folds and higher gaps develop in like manner as $\omega > 0$ increases. For instance, for $\omega = 4$, Fig. 3 reveals five folds and five gaps! Moreover, we claim that past all the folds, the steepness increases monotonically toward an extreme wave. Numerical computations (see [9], for instance) suggest that the extreme profile is single valued and exhibits a 120° corner at the crest, regardless of the value of the vorticity.

4.2 Touching Waves in the Infinite Vorticity Limit

The left panel of Fig. 4 displays the profiles of almost touching waves near the beginnings of the lowest gaps, and the right panel near the ends of the gaps, for four values of positive constant vorticity, in the (x, y) plane in the range $x \in [-2\pi, 2\pi]$. Touching is at $y = 0$. The profiles on the left resemble that in [25, Figure 4(b)].

At the beginnings of the gaps, we observe that s decreases monotonically toward ≈ 0.73 as $\omega \to \infty$ (see [8], for instance). Crapper [7] derived a remarkable formula

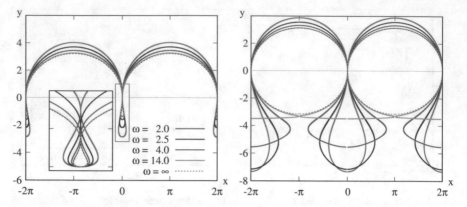

Fig. 4 On the left, touching waves at the beginnings of the lowest gaps for four values of vorticity. The dashed curved line is the limiting Crapper wave. On the right, touching waves at the ends of the gaps. The dashed curved line is a circle

of periodic capillary waves (in the absence of gravitational effects) in an irrotational flow of infinite depth, and calculated that $s \approx 0.73$ for the wave of greatest height. Moreover, the left panel reveals that, for instance, for $\omega = 14$, the profile of an almost touching wave is in excellent agreement with the limiting Crapper wave. Therefore, we claim that touching waves at the beginnings of the lowest gaps tend to the *limiting Crapper wave* as the value of positive constant vorticity increases indefinitely. It reveals a striking and surprising link between positive constant vorticity and capillarity!

At the ends of the gaps, on the other hand, we observe that $s \to 1$ as $\omega \to \infty$ (see [8], for instance). Teles da Silva and Peregrine [23], among others, numerically computed periodic waves in a constant vorticity flow in the absence of gravitational effects, and argued that a limiting wave has a circular shape made up of fluid in rigid body rotation (see also [25]). Moreover, the right panel reveals that, for instance, for $\omega = 14$, the profile of an almost touching wave is nearly circular. Therefore, we claim that touching waves at the ends of the lowest gaps tend to a *fluid disk in rigid body rotation* in the infinite vorticity limit. It is interesting to analytically explain the limiting Crapper wave and the circular vortex wave in the infinite vorticity limit.

Moreover in the left panel of Fig. 5 are the profiles of almost touching waves near the beginnings of the second gaps, and the right panel near the ends of the gaps, for three values of positive constant vorticity, in the (x, y) plane, where $x \in [-2\pi, 2\pi]$. The profile on the left for $\omega = 14$ resembles that in [25, Figure 5(c)], and the profile on the right resembles [25, Figure 6]. We may claim that touching waves at the beginnings of the second gaps tend to the circular vortex on top of the limiting Crapper wave as the value of positive constant vorticity increases indefinitely, whereas the circular vortex wave on top of itself at the ends of the gaps.

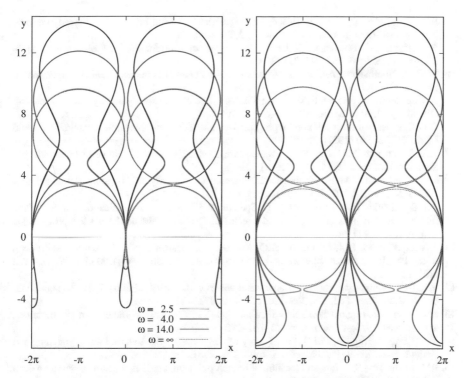

Fig. 5 On the left, touching waves at the beginnings of the second gaps for three values of vorticity (solid) and the circular vortex wave on top of the limiting Crapper wave (dashed). On the right, touching waves at the ends of the gaps (solid) and the circular vortex wave on top of itself (dashed)

We take matters further and claim that touching waves at the boundaries of higher gaps accommodate more circular vortices in like manner. See [9], for instance, for a profile nearly enclosing five circular vortices!

Acknowledgements VMH is supported by the US National Science Foundation under the Faculty Early Career Development (CAREER) Award DMS-1352597, and SD is supported by the National Science Foundation under DMS-1716822. VMH is grateful to the Erwin Schrödinger International Institute for Mathematics and Physics for its hospitality during the workshop Nonlinear Water Waves. This work was completed with the support of our TEX-pert.

References

1. K.I. Babenko, Some remarks on the theory of surface waves of finite amplitude. Soviet Math. Doklady **35**, 599–603 (1987) (See also loc. cit. 647–650)
2. B. Buffoni, J.F. Toland, *Analytic Theory of Global Bifurcation: An Introduction*. Princeton Series in Applied Mathematics (Princeton University Press, Princeton, 2003)

3. B. Buffoni, E.N. Dancer, J.F. Toland, The regularity and local bifurcation of steady periodic water waves. Arch. Ration. Mech. Anal. **152**, 207–240 (2002)
4. B. Buffoni, E.N. Dancer, J.F. Toland, The sub-harmonic bifurcation of Stokes waves. Arch. Ration. Mech. Anal. **152**, 241–271 (2002)
5. W. Choi, Nonlinear surface waves interacting with a linear shear current. Math. Comput. Simul. **80**, 101–110 (2009)
6. A. Constantin, W. Strauss, E. Varvaruca, Global bifurcation of steady gravity water waves with critical layers. Acta. Math. **217**, 195–262 (2016)
7. G.D. Crapper, An exact solution for progressive capillary waves of arbitrary amplitude. J. Fluid Mech. **2**, 532–540 (1957)
8. S.A. Dyachenko, V.M. Hur, Stokes waves with constant vorticity: I. Numerical computation. Stud. Appl. Math. **142**, 162–189 (2019)
9. S.A. Dyachenko, V.M. Hur, Stokes waves with constant vorticity: folds, gaps, and fluid bubbles. J. Fluid Mech. **878**, 502–521 (2019)
10. A.I. Dyachenko, E.A. Kuznetsov, M.D. Spector, V.E. Zakharov, Analytical description of the free surface dynamics of an ideal fluid (canonical formalism and conformal mapping). Phys. Lett. A **1**, 73–79 (1996)
11. S.A. Dyachenko, P.M. Lushnikov, A.O. Korotkevich, Branch cuts of Stokes waves on deep water. Part I: numerical solutions and Padé approximation. Stud. Appl. Math. **137**, 419–472 (2016)
12. V.H. Hur, Symmetry of steady periodic water waves with vorticity. Philos. Trans. R. Soc. Lond. Ser. A Math. Phys. Eng. Sci. **365**, 2203–2214 (2007)
13. M.S. Longuet-Higgins, Some new relations between Stokes's coefficients in the theory of gravity waves. J. Inst. Math. Appl. **22**, 261–273 (1978)
14. M.S. Longuet-Higgins, M.J.H. Fox, Theory of the almost-highest wave. Part 2. Matching and analytic extension. J. Fluid Mech. **85**, 769–786 (1978)
15. P.M. Lushnikov, Structure and location of branch point singularities for Stokes waves on deep water. J. Fluid Mech. **800**, 557–594 (2016)
16. P.M. Lushnikov, S.A. Dyachenko, D.A. Silantyev, New conformal mapping for adaptive resolving of the complex singularities of Stokes wave. Proc. R. Soc. A **473**, 20170198, 19 pp. (2017)
17. P.I. Plotnikov, Nonuniqueness of solutions of the problem of solitary waves and bifurcation of critical points of smooth functionals. Math. USSR Izvetiya **38**, 333–357 (1992)
18. R. Ribeiro, P.A. Milewski, A. Nachbin, Flow structure beneath rotational water waves with stagnation points. J. Fluid Mech. **812**, 792–814 (2017)
19. J.A. Simmen, P.G. Saffman, Steady deep-water waves on a linear shear current. Stud. Appl. Math. **73**, 35–57 (1985)
20. G.G. Stokes, On the theory of oscillatory waves. Trans. Camb. Phil. Soc. **8**, 441–445 (1847)
21. G.G. Stokes, Considerations relative to the greatest height of oscillatory irrotational waves which can be propagated without change of form, in *Mathematical and Physical Papers*, vol. I (Cambridge University Press, Cambridge, 1880), pp. 225–228
22. W.A. Strauss, Steady water waves. Bull. Amer. Math. Soc. (N.S.) **47**, 671–694 (2010)
23. A.F. Teles da Silva, D.H. Peregrine, Steep, steady surface waves on water of finite depth with constant vorticity. J. Fluid Mech. **195**, 281–302 (1988)
24. J.F. Toland, Stokes waves. Topol. Methods Nonlinear Anal. **7**, 1–48 (1996)
25. J.-M. Vanden-Broeck, Periodic waves with constant vorticity in water of infinite depth. IMA J. Appl. Math. **56**, 207–217 (1996)
26. A. Zygmund, *Trigonometric Series I & II*, corrected reprint (1968) of 2nd edn. (Cambridge University Press, Cambridge, 1959)

Integrable Models of Internal Gravity Water Waves Beneath a Flat Surface

Alan C. Compelli, Rossen I. Ivanov, and Tony Lyons

Abstract A two-layer fluid system separated by a pycnocline in the form of an internal wave is considered. The lower layer is bounded below by a flat bottom and the upper layer is bounded above by a flat surface. The fluids are incompressible and inviscid and Coriolis forces as well as currents are taken into consideration. A Hamiltonian formulation is presented and appropriate scaling leads to a KdV approximation. Additionally, considering the lower layer to be infinitely deep leads to a Benjamin–Ono approximation.

Keywords Internal waves · Currents · Nonlinear waves · Long waves · Hamiltonian systems · Solitons

Mathematics Subject Classification (2000) Primary: 35Q35, 35Q51, 35Q53; Secondary: 37K10

1 Introduction

The presented material provides a review of some well-known long wave models: the KdV and Benjamin–Ono approximations. The context is an oceanic fluid system comprising of two layers separated by an internal wave, created by a sharp density gradient, bounded above and below by a flat surface and flat seabed respectively.

A. C. Compelli
School of Mathematical Sciences, University College Cork, Cork, Ireland
e-mail: alan.compelli@ucc.ie

R. I. Ivanov (✉)
School of Mathematical Sciences, Technological University Dublin, Dublin, Ireland
e-mail: rossen.ivanov@tudublin.ie

T. Lyons
Department of Computing and Mathematics, Waterford Institute of Technology, Waterford, Ireland
e-mail: tlyons@wit.ie

© Springer Nature Switzerland AG 2019
D. Henry et al. (eds.), *Nonlinear Water Waves*, Tutorials, Schools, and Workshops in the Mathematical Sciences, https://doi.org/10.1007/978-3-030-33536-6_6

Many irrotational studies of both single layered and stratified systems such as [2–4, 17, 19, 20, 27, 28] have followed on from Zakharov's determination in [32] of a canonical Hamiltonian structure for a deep fluid with gravitational surface waves. The consideration of vorticity, however, is necessary for the inclusion of currents. The interaction of waves and currents have been examined for single layer systems in [10, 11, 14–16, 30, 31] and for stratified systems in [5–7, 12, 13].

2   The Set-Up

Consider a fluid system consisting of two domains as shown in Fig. 1. The lower medium is bounded underneath by a solid, stationary, impermeable layer of constant depth called the 'flatbed' at a depth h and the upper medium is bounded by a flat surface called the 'lid' at a height h_1. The physical reasoning is that the surface waves in the ocean have usually much smaller amplitudes in comparison to the internal waves. Typically h_1 may be of the order of hundreds of metres and the order of h may vary from hundreds of meters to several kilometers.

The system comprises of two separate fluids which have different densities due to different salinity levels and temperatures. Some prescribed flow has been generated by, perhaps, surface winds permeating downwards or due to tidal influences. However, at the interface the fluids do not mix and form a free common interface in the form of an internal wave. The wave is two-dimensional (in the x-y plane), propagating in the positive x-direction, due to the assumption that there is no lateral movement. This is a reasonable assumption for example, for oceanic waves of constant depth travelling along the equator [13, 22, 26]. The wave extends to infinity in both the positive and negative directions. The wave is characterised by the elevation function $\eta(x, t)$ with respect to the level $y = 0$. In other words the equation of the interface is

$$y = \eta(x, t). \tag{2.1}$$

Fig. 1 Set-up for the system

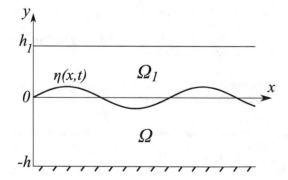

The mean value of η is taken to be zero for convenience,

$$\int_{\mathbb{R}} \eta(x,t)dx = 0, \quad \text{for all } t. \tag{2.2}$$

The system is assumed to be on the surface of the Earth, that is on a rotating solid body. The wave is acted upon by the restorative action of gravity. The Earth's centre of gravity is considered to be in the negative y-direction.

The domains Ω and Ω_1 are defined as

$$\Omega := \{(x, y) \in \mathbb{R}^2 : -h < y < \eta(x, t)\}$$

$$\text{and} \quad \Omega_1 := \{(x, y) \in \mathbb{R}^2 : \eta(x, t) < y < h_1\}.$$

Due to an assumption of incompressibility the constant densities are given by ρ and ρ_1 and stability is ensured by the assumption of immiscibility and that $\rho > \rho_1$.

The stream functions, ψ and ψ_1, are related to the velocity fields $\mathbf{u} = (u, v)$ and $\mathbf{u}_1 = (u_1, v_1)$ via the relations

$$u = \psi_y, \quad u_1 = \psi_{1,y}, \quad v = -\psi_x \quad \text{and} \quad v_1 = -\psi_{1,x} \tag{2.3}$$

due to the incompressibility assumption $\nabla \cdot \mathbf{u} = 0$, $\nabla \cdot \mathbf{u}_1 = 0$.

The velocity potentials, φ and φ_1, are introduced such that

$$u = \varphi_x + \gamma y, \quad u_1 = \varphi_{1,x} + \gamma_1 y, \quad v = \varphi_y \quad \text{and} \quad v_1 = \varphi_{1,y} \tag{2.4}$$

where γ and γ_1 are the constant vorticities, where the vorticities are defined as

$$\gamma = -v_x + u_y \quad \text{and} \quad \gamma_1 = -v_{1,x} + u_{1,y}. \tag{2.5}$$

This setup allows for modelling of an undercurrent, such as the Equatorial Undercurrent. A piecewise linear current profile can be represented by the velocity fields of the form (2.4), [12] by writing

$$u = \widetilde{\varphi}_x + \gamma y + \kappa, \quad u_1 = \widetilde{\varphi}_{1,x} + \gamma_1 y + \kappa_1, \quad v = \widetilde{\varphi}_y \quad \text{and} \quad v_1 = \widetilde{\varphi}_{1,y} \tag{2.6}$$

where κ and κ_1 are constants representing the current horizontal velocities at $y = 0$. The wave-only components have been separated out by introducing a tilde notation.

There is a harmonic conjugate relationship between ψ and $\widetilde{\varphi}$ (cf. [21, 25]) given by the complex analytic function

$$f(z) = \widetilde{\varphi}(x, y, t) + i\left(\psi(x, y, t) - \frac{1}{2}\gamma y^2 - \kappa y\right),$$

where $z = x + iy \in \Omega$, and similar for Ω_1. The fact that $f(z)$ is analytic in the corresponding domain allows the determination of the velocity potential $\widetilde{\varphi}(x, y, t)$ in Ω from its value $\phi(x, t)$ at the interface $y = \eta(x, t)$ (see (4.13) below, $\phi(x, t)$ can be expressed through the canonical Hamiltonian variables defined at the interface). Hence, the physical quantities in the body of the fluid can be determined from the variables at the interface as well.

We assume that the functions $\eta(x, t)$, $\widetilde{\varphi}(x, y, t)$ and $\widetilde{\varphi}_1(x, y, t)$ belong to the Schwartz class $S(\mathbb{R})$ (cf. [24]) with respect to x (for any y and t). The assumption of course implies that for large absolute values of x the internal wave attenuates, and is vanishing at infinity, and therefore

$$\lim_{|x|\to\infty} \eta(x, t) = \lim_{|x|\to\infty} \widetilde{\varphi}(x, y, t) = \lim_{|x|\to\infty} \widetilde{\varphi}_1(x, y, t) = 0. \tag{2.7}$$

Note that we have not specified the dynamics (the time-evolution) of our physical variables yet.

3   Governing Equations

The fluid velocities and the net forces per unit mass for the inviscid media under study are related through the Euler equations

$$\mathbf{u}_t + (\mathbf{u}.\nabla)\mathbf{u} = -\frac{1}{\rho}\nabla P + \mathbf{F} \quad \text{and} \quad \mathbf{u}_{1,t} + (\mathbf{u}_1.\nabla)\mathbf{u}_1 = -\frac{1}{\rho_1}\nabla P_1 + \mathbf{F}_1 \tag{3.1}$$

where

$$\mathbf{F} = 2\omega\nabla\psi \quad \text{and} \quad \mathbf{F}_1 = 2\omega\nabla\psi_1 \tag{3.2}$$

are the Coriolis forces per unit mass with ω being the rotational speed of the Earth. The pressures are given as static, dynamic and constant atmospheric pressure terms respectively, ρ and ρ_1 (due to the assumption of incompressibility) are the constant densities and g is the acceleration due to gravity. For the lower medium Ω,

$$P = \rho_1 g h_1 - \rho g y + p + p_{\text{atm}}, \tag{3.3}$$

and for the upper medium Ω_1 the total pressure is

$$P_1 = \rho_1 g(h_1 - y) + p_1 + p_{\text{atm}}. \tag{3.4}$$

The gradients of the dynamic pressures are given as

$$\nabla p = -\rho \nabla \left(\widetilde{\varphi}_t + \frac{1}{2} |\nabla \psi|^2 - (\gamma + 2\omega)\psi + gy \right)$$

$$\text{and} \quad \nabla p_1 = -\rho_1 \nabla \left(\widetilde{\varphi}_{1,t} + \frac{1}{2} |\nabla \psi_1|^2 - (\gamma_1 + 2\omega)\psi_1 + gy \right).$$

We can hence obtain the following Bernoulli condition at the interface (where $p = p_1$)

$$\rho \left((\widetilde{\varphi}_t)_c + \frac{1}{2} |\nabla \psi|^2_c - (\gamma + 2\omega)\chi + g\eta \right)$$
$$= \rho_1 \left((\widetilde{\varphi}_{1,t})_c + \frac{1}{2} |\nabla \psi_1|^2_c - (\gamma_1 + 2\omega)\chi_1 + g\eta \right) \quad (3.5)$$

where the subscript c signifies the evaluation at the common interface $y = \eta(x, t)$, $\chi = \psi(x, \eta, t)$ and $\chi_1 = \psi_1(x, \eta, t)$. Equation (3.5) will eventually produce the evolution of the quantity

$$\xi := \rho(\widetilde{\varphi})_c - \rho_1(\widetilde{\varphi}_1)_c$$

and this indicates that ξ can be chosen as a *momentum* variable in the Hamiltonian formulation of the problem. The obvious candidate for a counterpart *coordinate* variable is $\eta(x, t)$ and it evolves according to the so called *kinematic boundary condition* at the interface

$$\eta_t = v - u\eta_x = v_1 - u_1\eta_x. \quad (3.6)$$

This can be expressed in terms of the stream functions, using (2.3), as

$$\eta_t = -(\psi_x)_c - (\psi_y)_c \eta_x = -(\psi_{1,x})_c - (\psi_{1,y})_c \eta_x, \quad (3.7)$$

and in terms of the velocity potentials, using (2.6), as

$$\eta_t = (\widetilde{\varphi}_y)_c - \left((\widetilde{\varphi}_x)_c + \gamma\eta + \kappa \right)\eta_x = (\widetilde{\varphi}_{1,y})_c - \left((\widetilde{\varphi}_{1,x})_c + \gamma_1\eta + \kappa_1 \right)\eta_x. \quad (3.8)$$

The kinematic boundary condition at the bottom, requiring that there is no velocity component in the y-direction on the flat bed, is given by

$$\left(\widetilde{\varphi}(x, -h, t) \right)_y = 0 \quad \text{and} \quad \left(\psi(x, -h, t) \right)_x = 0 \quad (3.9)$$

and, additionally, there is a kinematic boundary condition at the top, requiring that there is no velocity component in the y-direction on the surface, given by

$$\left(\widetilde{\varphi}_1(x, h_1, t) \right)_y = 0 \quad \text{and} \quad \left(\psi_1(x, h_1, t) \right)_x = 0. \quad (3.10)$$

4 Hamiltonian Formulation

The functional H, which describes the total energy of the system, can be written as the sum of the kinetic, $\mathcal{K}$, and potential energy, $\mathcal{V}$ contributions. The potential part is

$$V(\eta) = \rho g \int_{\mathbb{R}} \int_{-h}^{\eta} y \, dy \, dx + \rho_1 g \int_{\mathbb{R}} \int_{\eta}^{h_1} y \, dy \, dx.$$

However, the potential energy is always measured from some reference value, e.g. $V(\eta = 0)$ which is the potential energy of the current (without wave motion). Therefore, the relevant part of the potential energy, contributing to the wave motion is

$$\mathcal{V}(\eta) = V(\eta) - V(0) = \rho g \int_{\mathbb{R}} \int_{0}^{\eta} y \, dy \, dx + \rho_1 g \int_{\mathbb{R}} \int_{\eta}^{0} y \, dy \, dx = \frac{1}{2}(\rho - \rho_1)g \int_{\mathbb{R}} \eta^2 \, dx.$$

In order to determine the kinetic energy of the wave motion, from the total kinetic energy of the fluid

$$\frac{1}{2}\rho \int_{\mathbb{R}} \int_{-h}^{\eta} (u^2 + v^2) \, dy \, dx + \frac{1}{2}\rho_1 \int_{\mathbb{R}} \int_{\eta}^{h_1} (u_1^2 + v_1^2) \, dy \, dx \tag{4.1}$$

one should subtract again the constant, but infinite kinetic energy of the current which is

$$\frac{1}{2}\rho \int_{\mathbb{R}} \int_{-h}^{0} (\gamma y + \kappa)^2 \, dy \, dx + \frac{1}{2}\rho_1 \int_{\mathbb{R}} \int_{0}^{h_1} (\gamma_1 y + \kappa_1)^2 \, dy \, dx. \tag{4.2}$$

In terms of the dependent variables $\eta(x, t)$, $\widetilde{\varphi}(x, t)$ and $\widetilde{\varphi}_1(x, t)$ this kinetic energy is

$$\mathcal{K}(\eta, \widetilde{\varphi}, \widetilde{\varphi}_1) = \frac{1}{2}\rho \int_{\mathbb{R}} \int_{-h}^{\eta} \left((\widetilde{\varphi}_x + \gamma y + \kappa)^2 + (\widetilde{\varphi}_y)^2 \right) dy \, dx - \frac{1}{2}\rho \int_{\mathbb{R}} \int_{-h}^{0} (\gamma y + \kappa)^2 \, dy \, dx$$

$$+ \frac{1}{2}\rho_1 \int_{\mathbb{R}} \int_{\eta}^{h_1} \left((\widetilde{\varphi}_{1,x} + \gamma_1 y + \kappa_1)^2 + (\widetilde{\varphi}_{1,y})^2 \right) dy \, dx - \frac{1}{2}\rho_1 \int_{\mathbb{R}} \int_{0}^{h_1} (\gamma_1 y + \kappa_1)^2 \, dy \, dx$$

$$= \frac{1}{2}\rho \int\limits_{\mathbb{R}} \int\limits_{-h}^{\eta} \left((\widetilde{\varphi}_x)^2 + (\widetilde{\varphi}_y)^2 + 2\widetilde{\varphi}_x(\gamma y + \kappa) \right) dy dx$$

$$+ \frac{1}{2}\rho_1 \int\limits_{\mathbb{R}} \int\limits_{\eta}^{h_1} \left((\widetilde{\varphi}_{1,x})^2 + (\widetilde{\varphi}_{1,y})^2 + 2\widetilde{\varphi}_{1,x}(\gamma_1 y + \kappa_1) \right) dy dx$$

$$+ \frac{1}{6}(\rho\gamma^2 - \rho_1\gamma_1^2) \int\limits_{\mathbb{R}} \eta^3 dx + \frac{1}{2}(\rho\gamma\kappa - \rho_1\gamma_1\kappa_1) \int\limits_{\mathbb{R}} \eta^2 dx. \quad (4.3)$$

The Hamiltonian is therefore

$$H(\eta, \widetilde{\varphi}, \widetilde{\varphi}_1) = \mathcal{K} + \mathcal{V} = \frac{1}{2}\rho \int\limits_{\mathbb{R}} \int\limits_{-h}^{\eta} \left((\widetilde{\varphi}_x)^2 + (\widetilde{\varphi}_y)^2 + 2\widetilde{\varphi}_x(\gamma y + \kappa) \right) dy dx$$

$$+ \frac{1}{2}\rho_1 \int\limits_{\mathbb{R}} \int\limits_{\eta}^{h_1} \left((\widetilde{\varphi}_{1,x})^2 + (\widetilde{\varphi}_{1,y})^2 + 2\widetilde{\varphi}_{1,x}(\gamma_1 y + \kappa_1) \right) dy dx$$

$$+ \frac{1}{6}(\rho\gamma^2 - \rho_1\gamma_1^2) \int\limits_{\mathbb{R}} \eta^3 dx + \frac{1}{2}\left((\rho\gamma\kappa - \rho_1\gamma_1\kappa_1) + (\rho - \rho_1)g \right) \int\limits_{\mathbb{R}} \eta^2 dx.$$

$$(4.4)$$

The Dirichlet–Neumann operators $G(\eta)$ and $G_1(\eta)$ are defined as [18]

$$G(\eta)\phi = (\widetilde{\varphi}_{\mathbf{n}})_c \sqrt{1 + \eta_x^2} \quad \text{and} \quad G_1(\eta)\phi_1 = (\widetilde{\varphi}_{1_{\mathbf{n}_1}})_c \sqrt{1 + \eta_x^2} \quad (4.5)$$

where $\mathbf{n}$ and $\mathbf{n}_1$ are the *unit* exterior normals, $\sqrt{1 + (\eta_x)^2}$ is a normalisation factor and

$$\phi(x, t) := (\widetilde{\varphi})_c = \widetilde{\varphi}(x, \eta(x, t), t) \quad \text{and} \quad \phi_1(x, t) := (\widetilde{\varphi}_1)_c = \widetilde{\varphi}_1(x, \eta(x, t), t)$$

$$(4.6)$$

have been introduced as the interface velocity potentials and also introduce the operator B [19] as

$$B := \rho G_1(\eta) + \rho_1 G(\eta). \quad (4.7)$$

Using the boundary conditions

$$\begin{cases} G(\eta)\phi = -\eta_x(\tilde{\varphi}_x)_c + (\tilde{\varphi}_y)_c = \eta_t + (\gamma\eta + \kappa)\eta_x, \\ G_1(\eta)\phi_1 = \eta_x(\tilde{\varphi}_{1,x})_c - (\tilde{\varphi}_{1,y})_c = -\eta_t - (\gamma_1\eta + \kappa_1)\eta_x \end{cases} \quad (4.8)$$

we get

$$G(\eta)\phi + G_1(\eta)\phi_1 = \mu \quad (4.9)$$

where

$$\mu := \big((\gamma - \gamma_1)\eta + (\kappa - \kappa_1)\big)\eta_x. \quad (4.10)$$

Introducing the *momentum* variable [2, 3]

$$\xi(x, t) = \rho\phi(x, t) - \rho_1\phi_1(x, t) \quad (4.11)$$

we can show that

$$B\phi = \rho_1 G(\eta)\phi + \rho G_1(\eta)\phi = \rho_1\mu + G_1(\eta)\xi \quad (4.12)$$

and thus

$$\begin{cases} \phi = B^{-1}\big(\rho_1\mu + G_1(\eta)\xi\big) \\ \phi_1 = B^{-1}\big(\rho\mu - G(\eta)\xi\big) \end{cases} \quad (4.13)$$

gives the explicit expression of ϕ and ϕ_1 in terms of η and ξ. Due to the initial assumptions on the velocity potentials, $\xi(x, t)$ is a Schwartz class $\mathcal{S}(\mathbb{R})$ function in x (for any t).

Usually there is no jump in the current velocity, hence in what follows we take $\kappa = \kappa_1$. The Hamiltonian of the system can be expressed in terms of variables defined on the interface only, η and ξ:

$$H(\eta, \xi) = \frac{1}{2}\int_{\mathbb{R}} \xi G(\eta) B^{-1} G_1(\eta)\xi \, dx - \frac{1}{2}\rho\rho_1(\gamma - \gamma_1)^2 \int_{\mathbb{R}} \eta\eta_x B^{-1}\eta\eta_x dx$$

$$-\gamma \int_{\mathbb{R}} \xi\eta\eta_x dx - \kappa \int_{\mathbb{R}} \xi\eta_x dx + \rho_1(\gamma - \gamma_1) \int_{\mathbb{R}} \eta\eta_x B^{-1} G(\eta)\xi \, dx + \frac{1}{6}(\rho\gamma^2 - \rho_1\gamma_1^2) \int_{\mathbb{R}} \eta^3 dx$$

$$+ \frac{1}{2}\big((\rho\gamma - \rho_1\gamma_1)\kappa + g(\rho - \rho_1)\big) \int_{\mathbb{R}} \eta^2 dx. \quad (4.14)$$

It is a natural physical fact that there is no flow through the common interface and therefore the stream functions $\chi = \psi(x, \eta, t)$ and $\chi_1 = \psi_1(x, \eta, t)$ at the interface coincide,

$$\chi = \chi_1 = -\int_{-\infty}^{x} \eta_t(x', t)dx' = -\partial_x^{-1}\eta_t \qquad (4.15)$$

noting that due to (3.7)

$$\frac{d}{dx}\psi(x, \eta, t) = \psi_x + \psi_y(x, \eta, t)\eta_x = -\eta_t.$$

By evaluating the variations of the Hamiltonian one can show that (3.8) and (3.5) can be written in the form of a non-canonical Hamiltonian system [16]

$$\eta_t = \frac{\delta H}{\delta \xi} \quad \text{and} \quad \xi_t = -\frac{\delta H}{\delta \eta} + \Gamma\chi = -\frac{\delta H}{\delta \eta} - \Gamma\partial_x^{-1}\eta_t, \qquad (4.16)$$

where

$$\Gamma := \rho\gamma - \rho_1\gamma_1 + 2\omega(\rho - \rho_1) \qquad (4.17)$$

is a constant. Canonical equations of motion can be achieved by transforming the velocity potential at the interface, ξ, to a new variable, ζ, via the transformation (*cf.* [31])

$$\xi \quad \rightarrow \quad \zeta = \xi + \frac{\Gamma}{2}\int_{-\infty}^{x} \eta(x', t)dx, \qquad (4.18)$$

and due to (2.2) the variable $\zeta \in S(\mathbb{R})$ (for any t). For our further convenience however Eq. (4.16) will be written in terms of the variable

$$u = \xi_x$$

and hence for a Hamiltonian in terms of u and η

$$\eta_t = -\left(\frac{\delta H}{\delta u}\right)_x \quad \text{and} \quad u_t = -\left(\frac{\delta H}{\delta \eta}\right)_x - \Gamma\eta_t. \qquad (4.19)$$

5 Expanding the Dirichlet–Neumann Operators

The Dirichlet–Neumann operators can be expanded in terms of powers of η as

$$G(\eta) = \sum_{j=0}^{\infty} G^{(j)}(\eta) \text{ and } G_1(\eta) = \sum_{j=0}^{\infty} G_1^{(j)}(\eta), \qquad (5.1)$$

where $G^{(j)}(\eta)$ is a homogeneous expression in η of degree j, that is $G^{(j)}(b\eta) = b^j G^{(j)}(\eta)$ for any constant b. The explicit expansion is [19]

$$G(\eta) = DT(D) + D\eta D - DT(D)\eta DT(D) + \mathcal{O}(\eta^2) \tag{5.2}$$

$$\text{and} \quad G_1(\eta) = DT_1(D) - D\eta D + DT_1(D)\eta DT_1(D) + \mathcal{O}(\eta^2) \tag{5.3}$$

where

$$D := -i\partial_x \tag{5.4}$$

is a differential operator and

$$T(D) := \tanh(hD) \quad \text{and} \quad T_1(D) := \tanh(h_1 D) \tag{5.5}$$

have been introduced.

The operator B, as defined in (4.7), which is a function of the Dirichlet–Neumann operators, can therefore be expressed as

$$B = \rho \sum_{j=0}^{\infty} G_1^{(j)}(\eta) + \rho_1 \sum_{j=0}^{\infty} G^{(j)}(\eta).$$

It is noted that the leading (zeroth order in η) term in the expansion of B^{-1}, represented by $[B^{-1}]^{(0)}$, is

$$[B^{-1}]^{(0)} = \frac{1}{\rho DT_1(D) + \rho_1 DT(D)}. \tag{5.6}$$

6 Approximations

6.1 The KdV Approximation

A KdV-type approximation will be derived (*cf.* [8]). This family of equations are characterised as having weakly nonlinear and dispersive components.

Small parameters associated to the physical scales

$$\varepsilon = \frac{a}{h_1} \quad \text{and} \quad \delta = \frac{h_1}{\lambda} \tag{6.1}$$

are introduced where λ is the wavelength of the internal wave and a is the average wave amplitude. Indeed, $\delta \ll 1$ is small for long waves $\lambda \gg h_1$. This approximation therefore is for the long-wave regime. The quantity $h_1 k$ where $k = 2\pi/\lambda$ is the wave number is therefore scaled as

$$\mathcal{O}(h_1 k) = \delta,$$

and therefore for the operator D (which on monochromatic waves has an eigenvalue equal to the wave number) clearly

$$\mathcal{O}(h_1 D) = \delta. \tag{6.2}$$

To keep track of the order of the variables we replace $h_1 D$ with $\delta h_1 D$ and further assume that $h_1 D$ itself is of order 1. Since h and h_1 are fixed constants, then their ratio is of order 1. The wave elevation function is scaled according to

$$\eta \to \varepsilon \eta. \tag{6.3}$$

It can be shown as in [8] that the scaling of ξ, leading to the KdV approximation is

$$\xi \to \delta \xi. \tag{6.4}$$

The expansion of the Dirichlet–Neumann operators, given in (5.2) and (5.3), can be scaled as

$$G(\eta) \to \delta\big(D\tanh(\delta h D)\big) + \varepsilon\delta^2\big(D\eta D - D\tanh(\delta h D)\eta D\tanh(\delta h D)\big) + \mathcal{O}(\varepsilon^2\delta^4)$$

$$G_1(\eta) \to \delta\big(D\tanh(\delta h_1 D)\big) - \varepsilon\delta^2\big(D\eta D - D\tanh(\delta h_1 D)\eta D\tanh(\delta h_1 D)\big)$$
$$+ \mathcal{O}(\varepsilon^2\delta^4).$$

Using the expansion for the hyperbolic tangent the Dirichlet–Neumann operators can be represented as

$$G(\eta) = \delta^2\Big(h D^2 + \varepsilon D\eta D\Big) - \delta^4\Big(\frac{1}{3}h^3 D^4 + \varepsilon h^2 D^2\eta D^2\Big)$$

$$+ \delta^6\Big(\frac{2}{15}h^5 D^6\Big) + \mathcal{O}(\delta^8, \varepsilon\delta^6, \varepsilon^2\delta^4) \tag{6.5}$$

and

$$G_1(\eta) = \delta^2\Big(h_1 D^2 - \varepsilon D\eta D\Big) + \delta^4\Big(-\frac{1}{3}h_1^3 D^4 + \varepsilon h_1^2 D^2\eta D^2\Big)$$

$$+ \delta^6\Big(\frac{2}{15}h_1^5 D^6\Big) + \mathcal{O}(\delta^8, \varepsilon\delta^6, \varepsilon^2\delta^4). \tag{6.6}$$

and so the inverse of the operator B is given by

$$
\begin{aligned}
B^{-1} = {} & \frac{1}{\delta^2(\rho_1 h + \rho h_1)} D^{-1} \Bigg\{ 1 - \varepsilon \frac{\rho_1 - \rho}{\rho_1 h + \rho h_1} \eta + \varepsilon^2 \frac{(\rho_1 - \rho)^2}{(\rho_1 h + \rho h_1)^2} \eta^2 \\
& + \delta^2 \Bigg(\frac{1}{3} \frac{\rho_1 h^3 + \rho h_1^3}{\rho_1 h + \rho h_1} D^2 - \frac{1}{3} \varepsilon \frac{(\rho_1 - \rho)(\rho_1 h^3 + \rho h_1^3)}{(\rho_1 h + \rho h_1)^2} \eta D^2 \\
& - \frac{1}{3} \varepsilon \frac{(\rho_1 - \rho)(\rho_1 h^3 + \rho h_1^3)}{(\rho_1 h + \rho h_1)^2} D^2 \eta + \varepsilon \frac{\rho_1 h^2 - \rho h_1^2}{\rho_1 h + \rho h_1} D \eta D \Bigg) \\
& - \delta^4 \Bigg(\frac{2}{15} \frac{\rho_1 h^5 + \rho h_1^5}{\rho_1 h + \rho h_1} D^4 - \frac{1}{9} \frac{(\rho_1 h^3 + \rho h_1^3)^2}{(\rho_1 h + \rho h_1)^2} D^4 \Bigg) + \mathcal{O}(\delta^6, \varepsilon \delta^4, \varepsilon^2 \delta^2, \varepsilon^3) \Bigg\} D^{-1}.
\end{aligned}
\tag{6.7}
$$

By assuming that ε and δ^2 are of the same order, so as to permit a balancing between nonlinearity and dispersion, the Hamiltonian to $\mathcal{O}(\delta^6)$ is therefore

$$
\begin{aligned}
H(\eta, \xi) = {} & \frac{1}{2} \delta^4 \alpha_1 \int_{\mathbb{R}} \xi D^2 \xi \, dx + \frac{1}{2} \delta^6 \alpha_3 \int_{\mathbb{R}} \xi D \eta D \xi \, dx - \frac{1}{2} \delta^6 \alpha_2 \int_{\mathbb{R}} \xi D^4 \xi \, dx \\
& - \delta^4 \kappa \int_{\mathbb{R}} \xi \eta_x \, dx - \delta^6 \alpha_4 \int_{\mathbb{R}} \xi \eta \eta_x \, dx + \frac{1}{6} \delta^6 \alpha_6 \int_{\mathbb{R}} \eta^3 \, dx + \frac{1}{2} \delta^4 \alpha_5 \int_{\mathbb{R}} \eta^2 \, dx
\end{aligned}
\tag{6.8}
$$

or

$$
\begin{aligned}
H(\eta, \mathfrak{u}) = {} & \frac{1}{2} \delta^4 \alpha_1 \int_{\mathbb{R}} \mathfrak{u}^2 \, dx + \frac{1}{2} \delta^6 \alpha_3 \int_{\mathbb{R}} \eta \mathfrak{u}^2 \, dx - \frac{1}{2} \delta^6 \alpha_2 \int_{\mathbb{R}} \mathfrak{u}_x^2 \, dx \\
& + \delta^4 \kappa \int_{\mathbb{R}} \eta \mathfrak{u} \, dx + \delta^6 \frac{1}{2} \alpha_4 \int_{\mathbb{R}} \mathfrak{u} \eta^2 \, dx + \frac{1}{6} \delta^6 \alpha_6 \int_{\mathbb{R}} \eta^3 \, dx + \frac{1}{2} \delta^4 \alpha_5 \int_{\mathbb{R}} \eta^2 \, dx
\end{aligned}
\tag{6.9}
$$

where the following constants have been introduced

$$
\alpha_1 = \frac{h h_1}{\rho_1 h + \rho h_1},
\tag{6.10}
$$

$$
\alpha_2 = \frac{1}{3} \frac{h^2 h_1^2 (\rho_1 h_1 + \rho h)}{(\rho_1 h + \rho h_1)^2},
\tag{6.11}
$$

$$
\alpha_3 = \frac{\rho h_1^2 - \rho_1 h^2}{(\rho_1 h + \rho h_1)^2},
\tag{6.12}
$$

$$
\alpha_4 = \frac{\gamma \rho h_1 + \gamma_1 \rho_1 h}{\rho_1 h + \rho h_1},
\tag{6.13}
$$

$$\alpha_5 = (\rho\gamma - \rho_1\gamma_1)\kappa + g(\rho - \rho_1), \tag{6.14}$$

$$\alpha_6 = \rho\gamma^2 - \rho_1\gamma_1^2. \tag{6.15}$$

The equations of motion (4.19) are now written in terms of η and u as

$$\eta_t + \kappa\eta_x + \alpha_1 u_x + \delta^2\alpha_3(u\eta)_x + \delta^2\alpha_2 u_{xxx} + \delta^2\alpha_4\eta\eta_x = 0 \tag{6.16}$$

and $\quad u_t + \kappa u_x + \delta^2\alpha_3 u u_x + \delta^2\alpha_4(u\eta)_x + \delta^2\alpha_6\eta\eta_x + \alpha_5\eta_x + \Gamma\eta_t = 0,$
$$\tag{6.17}$$

with an appropriate scaling of t. Noting the assumption that $g \gg 2\omega\kappa$ and introducing a Galilean shift

$$X = x - \kappa t, \quad T = t, \quad \partial_X = \partial_x \quad \text{and} \quad \partial_T = \partial_t + \kappa\partial_x \tag{6.18}$$

the equations of motion can be written as

$$\eta_T + \alpha_1 u_X + \delta^2\big(\alpha_2 u_{XXX} + \alpha_3(u\eta)_X + \alpha_4\eta\eta_X\big) = 0 \tag{6.19}$$

and $\quad u_T - \Gamma\alpha_1 u_X + g(\rho - \rho_1)\eta_X + \delta^2\big(-\Gamma\alpha_2 u_{XXX}$

$$+ \alpha_3 u u_X + \alpha_4(u\eta)_X - \Gamma\alpha_3(u\eta)_X + \alpha_6\eta\eta_X - \Gamma\alpha_4\eta\eta_X\big) = 0. \tag{6.20}$$

The leading order linearised equations are therefore

$$\eta_T + \alpha_1 u_X = 0 \tag{6.21}$$

$$\text{and } u_T - \Gamma\alpha_1 u_X + g(\rho - \rho_1)\eta_X = 0. \tag{6.22}$$

The variables, η and u can be represented as

$$\eta(X, T) = \eta_0 e^{i(kX - \Omega(k)T)} \tag{6.23}$$

$$\text{and } u(X, T) = u_0 e^{i(kX - \Omega(k)T)}. \tag{6.24}$$

Noting that the wave number, angular frequency and wave speed are related via $c(k) = \Omega(k)/k$ means it can be written that

$$-ick\eta + i\alpha_1 ku = 0 \tag{6.25}$$

$$\text{and } -icku + ig(\rho - \rho_1)k\eta - i\Gamma\alpha_1 ku = 0. \tag{6.26}$$

This has solutions for observers moving with the flow as

$$c = \frac{1}{2}\left(-\Gamma\alpha_1 \pm \sqrt{\alpha_1^2\Gamma^2 + 4\alpha_1 g(\rho - \rho_1)}\right). \tag{6.27}$$

From (6.25) in the leading order $u = \frac{c}{\alpha_1}\eta$. Considering a relation that goes to the next order

$$u = \frac{c}{\alpha_1}\eta + \delta^2\left(\sigma\eta_{XX} + \mu\eta^2\right) \tag{6.28}$$

for some constants μ and σ we can exclude u from the system (6.19)–(6.20) and write both equations in terms of η. Of course they should coincide for the special choice of the constants μ and σ which is

$$\sigma = -\frac{c\alpha_2(c + \Gamma\alpha_1)}{\alpha_1^2(2c + \Gamma\alpha_1)} \tag{6.29}$$

and

$$\mu = \frac{\alpha_1\alpha_4(c - \Gamma\alpha_1) - \alpha_3 c(c + 2\Gamma\alpha_1) + \alpha_1^2\alpha_6}{2\alpha_1^2(2c + \Gamma\alpha_1)} \tag{6.30}$$

giving the KdV equation

$$\eta_T + c\eta_X + \delta^2\left(\frac{c^2\alpha_2}{\alpha_1(2c + \Gamma\alpha_1)}\right)\eta_{XXX} + \delta^2\left(\frac{\alpha_1^2\alpha_6 + 3\alpha_3 c^2 + 3\alpha_1\alpha_4 c}{\alpha_1(2c + \Gamma\alpha_1)}\right)\eta\eta_X = 0. \tag{6.31}$$

Recalling the constants (6.10)–(6.15) when $\gamma = \gamma_1 = \omega = 0$ this becomes

$$\eta_T + c\eta_X + \delta^2\frac{chh_1(\rho_1 h_1 + \rho h)}{6(\rho_1 h + \rho h_1)}\eta_{XXX} + \frac{3}{2}\delta^2 c\frac{\rho h_1^2 - \rho_1 h^2}{hh_1(\rho_1 h + \rho h_1)}\eta\eta_X = 0, \tag{6.32}$$

where

$$c = \pm\sqrt{\frac{hh_1(\rho - \rho_1)g}{\rho_1 h + \rho h_1}} = \pm\sqrt{\frac{(\rho - \rho_1)g}{\rho_1/h_1 + \rho/h}}.$$

In the case $h \to \infty$ we have

$$c_\infty = \pm\sqrt{\frac{h_1(\rho - \rho_1)g}{\rho_1}}. \tag{6.33}$$

Next, we recall fact that the canonical KdV equation

$$E_T + E_{XXX} + 6EE_X = 0 \qquad (6.34)$$

has a one-soliton solution

$$E(X, T) = 2v^2 \text{sech}^2 v(X - 4v^2 T - X_0)$$

where v, X_0 are constants, related to the soliton's initial position and velocity.

Let us now introduce

$$\mathcal{A} = \delta^2 \frac{\alpha_1^2 \alpha_6 + 3\alpha_3 c^2 + 3\alpha_1 \alpha_4 c}{\alpha_1 (2c + \Gamma \alpha_1)}$$

$$\mathcal{B} = \delta^2 \frac{c^2 \alpha_2}{\alpha_1 (2c + \Gamma \alpha_1)}$$

and rescale the variables

$$\eta = \alpha E, \qquad X \to \beta X, \qquad T \to \beta T$$

in order to match the coefficients of (6.34). This gives $\alpha = 6\beta^2 \mathcal{B}/\mathcal{A}$. Applying further a Galilean shift we obtain the one-soliton solution of (6.31) as

$$\eta(X, T) = \frac{12\mathcal{B}}{\mathcal{A}} v^2 \beta^2 \text{sech}^2 \left(v\beta(X - X_0 - (c + 4v^2 \beta^2 \mathcal{B})T) \right).$$

Introducing the constant $K = v\beta$ which has a dimensionality $(\text{length})^{-1}$ and the meaning of an analogue of a wave number, the above formula becomes

$$\eta(X, T) = \frac{12\mathcal{B}}{\mathcal{A}} K^2 \text{sech}^2 \left(K(X - X_0 - (c + 4K^2 \mathcal{B})T) \right). \qquad (6.35)$$

The maximal amplitude of the solitary wave is therefore

$$\eta_0 = \frac{12\mathcal{B}}{\mathcal{A}} K^2$$

and it is related to the constant K. The propagation speed is

$$V = c + 4K^2 \mathcal{B}$$

which is represented from the component of the leading order linear wave c and the soliton speed $4K^2 \mathcal{B}$ which is proportional to the amplitude η_0 due to the K^2 factor.

Let us now analyse the irrotational case where

$$\eta_0 = \frac{4K^2 h^2 h_1^2 (\rho_1 h_1 + \rho h)}{3(\rho h_1^2 - \rho_1 h^2)}.$$

Since ρ and ρ_1 are very close, and usually h is much bigger than h_1, then $\eta_0 < 0$ and the soliton is a depression wave. The velocity is

$$V = c \left(1 + \delta^2 \frac{2}{3} K^2 h h_1 \right) = \pm \sqrt{\frac{h h_1 (\rho - \rho_1) g}{\rho_1 h + \rho h_1}} \left(1 + \delta^2 \frac{2}{3} K^2 h h_1 \frac{\rho_1 h_1 + \rho h}{\rho_1 h + \rho h_1} \right).$$

The plus and minus signs are for the right and left running waves respectively. Therefore the bigger wave travels faster.

6.2   The Benjamin–Ono Approximation

For the Benjamin–Ono approximation we consider the system with an infinitely deep lower layer $h \to \infty$ (*cf.* [9]). The Hamiltonian is (4.14) with the following scaling

$$\eta \to \delta\eta, \quad \xi \to \xi \quad \text{and} \quad D \to \delta D. \tag{6.36}$$

The Dirichlet–Neumann operators, given in (5.2) and (5.3), can be expanded, taking into account that

$$\lim_{h \to \infty} \tanh(hD) = \text{sgn}(D), \qquad \lim_{h \to \infty} D \tanh(hD) = |D|.$$

In order to explain the meaning of $|D|$, we introduce the Fourier transform

$$\hat{u}(k) := \mathcal{F}\{u(x)\}(k), \qquad u(x) = \mathcal{F}^{-1}\{\hat{u}(k)\}(x).$$

Then

$$|D|u(x) := \mathcal{F}^{-1}\{|k|\hat{u}(k)\}(x)$$

and similarly

$$\text{sgn}(D)u(x) := \mathcal{F}^{-1}\{\text{sgn}(k)\hat{u}(k)\}(x).$$

There is a relation between the Hilbert transform, $\mathcal{H}$

$$\mathcal{H}\{u\}(x) := \text{P.V.} \frac{1}{\pi} \int_{-\infty}^{\infty} \frac{u(x')dx'}{x - x'}$$

and the Fourier transforms, namely

$$\mathcal{F}\{\mathcal{H}\{u\}(x)\}(k) = -i\,\mathrm{sgn}(k)\hat{u}(k)$$

or

$$\mathcal{H}\{u\}(x) = -i\,\mathcal{F}^{-1}\{\mathrm{sgn}(k)\hat{u}(k)\}(x).$$

Hence

$$\mathcal{H}\{Du\}(x) = -i\,\mathcal{F}^{-1}\{|k|\hat{u}(k)\}(x) = -i|D|u(x),$$

or

$$|D| = i\mathcal{H}D = \mathcal{H}\partial_x.$$

The expansion is

$$G(\eta) = \delta|D| + \delta^3\big(D\eta D - |D|\eta|D|\big) + \mathcal{O}(\delta^5)$$
$$\text{and}\quad G_1(\eta) = \delta D\tanh(\delta h_1 D)$$
$$-\delta^3\big(D\eta D - D\tanh(\delta h_1 D)\eta D\tanh(\delta h_1 D)\big) + \mathcal{O}(\delta^6)$$

noting from [19] that the leading term for the infinite lower layer is $|D|$. Using the expansion for the tanh, the Dirichlet–Neumann operators can be represented further as

$$G(\eta) = \delta|D| + \delta^3 D\eta D - \delta^3|D|\eta|D| + \mathcal{O}(\delta^5)$$
$$\text{and}\quad G_1(\eta) = \delta^2 h_1 D^2 - \delta^3 D\eta D + \mathcal{O}(\delta^4)$$

and so the inverse of the operator B is given by

$$B^{-1} = \frac{1}{\delta\rho_1}|D|D^{-1}\left\{1 - \delta\frac{\rho}{\rho_1}h_1|D| + \mathcal{O}(\delta^2)\right\}D^{-1}.$$

The Hamiltonian can therefore be written, using components of the expanded operators as (see the notations (6.14) and (6.15))

$$H(\eta, \xi) = \frac{1}{2}\delta^2\frac{h_1}{\rho_1}\int_{\mathbb{R}}\xi D^2\xi\,dx - \frac{1}{2}\delta^3\frac{h_1^2\rho}{\rho_1^2}\int_{\mathbb{R}}\xi|D|D^2\xi\,dx - \frac{1}{2}\delta^3\frac{1}{\rho_1}\int_{\mathbb{R}}\xi D\eta D\xi\,dx$$

$$- \delta^3\gamma_1\int_{\mathbb{R}}\xi\eta\eta_x dx - \delta^2\kappa\int_{\mathbb{R}}\xi\eta_x dx + \frac{1}{6}\delta^3\alpha_6\int_{\mathbb{R}}\eta^3 dx + \frac{1}{2}\delta^2\alpha_5\int_{\mathbb{R}}\eta^2 dx + \mathcal{O}(\delta^4)$$

$$(6.37)$$

and in terms of η, u

$$H(\eta, u) = \frac{1}{2}\delta^2\frac{h_1}{\rho_1}\int_{\mathbb{R}} u^2\,dx - \frac{1}{2}\delta^3\frac{h_1^2\rho}{\rho_1^2}\int_{\mathbb{R}} u|D|u\,dx - \frac{1}{2}\delta^3\frac{1}{\rho_1}\int_{\mathbb{R}} \eta u^2\,dx$$

$$+ \delta^3\frac{\gamma_1}{2}\int_{\mathbb{R}} u\eta^2\,dx + \delta^2\kappa\int_{\mathbb{R}} u\eta\,dx + \frac{1}{6}\delta^3\alpha_6\int_{\mathbb{R}} \eta^3\,dx + \frac{1}{2}\delta^2\alpha_5\int_{\mathbb{R}} \eta^2\,dx + \mathcal{O}(\delta^4).$$

$$(6.38)$$

The equations of motion (4.19) are now written in terms of η and u as

$$\eta_t + \kappa\eta_x + \frac{h_1}{\rho_1}u_x - \delta\frac{h_1^2\rho}{\rho_1^2}|D|u_x - \delta\frac{1}{\rho_1}(\eta u)_x + \delta\gamma_1\eta\eta_x = 0 \qquad (6.39)$$

and $\quad u_t + \kappa u_x - \delta\frac{1}{\rho_1}uu_x + \delta\gamma_1(\eta u)_x + \delta\alpha_6\eta\eta_x + \alpha_5\eta_x + \Gamma\eta_t = 0.$ $\qquad$ (6.40)

Again we perform the Galilean shift (6.18) noting that $g \gg 2\omega\kappa$ and $\alpha_5 - \Gamma\kappa \approx g(\rho - \rho_1)$ to obtain

$$\eta_T + \frac{h_1}{\rho_1}u_X - \delta\frac{h_1^2\rho}{\rho_1^2}|D|u_X - \delta\frac{1}{\rho_1}(\eta u)_X + \delta\gamma_1\eta\eta_X = 0 \qquad (6.41)$$

and $\quad u_T - \delta\frac{1}{\rho_1}uu_X + \delta\gamma_1(\eta u)_X + \delta\alpha_6\eta\eta_X + g(\rho - \rho_1)\eta_X + \Gamma\eta_T = 0.$ $\quad$ (6.42)

In the leading order

$$\eta_T = -\frac{h_1}{\rho_1}u_X \quad \text{and} \quad u_T = -g(\rho - \rho_1)\eta_X - \Gamma\eta_T.$$

Again using exponential representations (6.23) the above equations give

$$-c\eta = -\frac{h_1}{\rho_1}u \qquad (6.43)$$

and $\quad -cu = \left(-g(\rho - \rho_1) + c\Gamma\right)\eta.$ $\qquad$ (6.44)

This gives an equation $c^2 = -h_1\left(-g(\rho - \rho_1) + c\Gamma\right)/\rho_1$ with solutions

$$c = -\frac{h_1}{2\rho_1}\Gamma \pm \frac{1}{2}\sqrt{\frac{h_1^2}{\rho_1^2}\Gamma^2 + 4\frac{h_1}{\rho_1}g(\rho - \rho_1)}. \qquad (6.45)$$

Considering an expansion of the type of (6.28)

$$u = \frac{\rho_1}{h_1} c\eta + \delta\alpha\eta^2 + \delta\beta|D|\eta,$$

we can determine that

$$\alpha = \frac{\rho_1(\rho_1 c^2 + 2h_1\Gamma c - \gamma_1 h_1^2\Gamma + \rho_1\gamma_1 h_1 c + h_1^2\alpha_6)}{2h_1^2(2\rho_1 c + h_1\Gamma)} \tag{6.46}$$

and

$$\beta = \frac{\rho(\rho_1 c^2 + h_1\Gamma c)}{2\rho_1 c + h_1\Gamma}. \tag{6.47}$$

The equation for η is therefore given by

$$\eta_T + c\eta_X - \delta\frac{\rho h_1 c^2}{2\rho_1 c + h_1\Gamma}|\partial_X|\eta_X + \delta\frac{-3\rho_1 c^2 + 3\rho_1\gamma_1 h_1 c + h_1^2\alpha_6}{h_1(2\rho_1 c + h_1\Gamma)}\eta\eta_x = 0. \tag{6.48}$$

The obtained equation is the well known Benjamin–Ono (BO) equation [1, 29] which is an integrable equation whose solutions can be obtained by the Inverse Scattering method [23].

The Benjamin–Ono equation in the irrotational case ($\gamma = \gamma_1 = \omega = 0$, $\alpha_6 = \Gamma = 0$) becomes (cf. [5])

$$\eta_t + c\eta_x - \frac{1}{2}\delta\frac{\rho h_1 c}{\rho_1}|D|\eta_x - \frac{3}{2}\delta\frac{c}{h_1}\eta\eta_x = 0, \tag{6.49}$$

where, from (6.45)

$$c = \pm\sqrt{\frac{h_1}{\rho_1}g(\rho - \rho_1)}.$$

This wavespeed of course coincides with (6.33).

The BO equation in the form

$$\eta_T + c\eta_X + A\eta\eta_x + B|\partial_X|\eta_X = 0 \tag{6.50}$$

has a one-soliton solution

$$\eta(X, T) = \frac{\eta_0}{1 + \left(\frac{A\eta_0}{4B}\right)^2\left[X - X_0 - \left(c + \frac{1}{4}A\eta_0\right)T\right]^2} \tag{6.51}$$

where the amplitude η_0 and the initial displacement X_0 are arbitrary constants. From (6.48) for the internal wave equation

$$\mathcal{A} := \delta \frac{-3\rho_1 c^2 + 3\rho_1 \gamma_1 h_1 c + h_1^2 \alpha_6}{h_1(2\rho_1 c + h_1 \Gamma)} \tag{6.52}$$

and

$$\mathcal{B} := \delta \frac{\rho h_1 c^2}{2\rho_1 c + h_1 \Gamma}. \tag{6.53}$$

We note that (6.51) shows that the wavespeed of the soliton $c + \frac{1}{4}\mathcal{A}\eta_0$ depends on its amplitude η_0 and on the parameters of the system.

7 Discussion

The illustrative one-soliton solutions of the KdV (6.35) and the BO equation (6.51) suffers, however, from the following disadvantages. First, the BO soliton is not in the Schwartz class in the x-variable, which is not a very serious disadvantage from the physical point of view. Second, the assumption (2.2) for η is violated since for the one-soliton solutions have finite "mass" proportional to $\int_{\mathbb{R}} \eta(X, T)dX$, which for the KdV model is $24\mathcal{B}K/\mathcal{A}$ and for the BO model is $\pi\mathcal{B}/\mathcal{A}$. One can argue again that this does not change the physical setup. Indeed, the average value of η would be

$$\langle \eta \rangle = \frac{\int_{\mathbb{R}} \eta(X, T)dX}{\int_{\mathbb{R}} dX} = 0$$

since the nominator is finite and the denominator is infinite. We note also that the "mass" $\int_{\mathbb{R}} \eta(X, T)dX$, is always a conserved quantity due to (4.19). Therefore the extra condition (2.2) can be properly relaxed, allowing for solitary waves with a finite "mass".

Acknowledgements The authors are grateful to the Erwin Schrödinger International Institute for Mathematics and Physics (ESI), Vienna (Austria) for the opportunity to participate in the workshop *Nonlinear Water Waves—an Interdisciplinary Interface*, 2017 where a significant part of this work has been accomplished. AC is also funded by SFI grant 13/CDA/2117.

References

1. T.B. Benjamin, Internal waves of permanent form in fluids of great depth. J. Fluid Mech. **29**, 559–562 (1967)

2. T.B. Benjamin, T.J. Bridges, Reappraisal of the Kelvin-Helmholtz problem. Part 1. Hamiltonian structure. J. Fluid Mech. **333**, 301–325 (1997)
3. T.B. Benjamin, T.J. Bridges, Reappraisal of the Kelvin-Helmholtz problem. Part 2. Interaction of the Kelvin-Helmholtz, superharmonic and Benjamin-Feir instabilities. J. Fluid Mech. **333**, 327–373 (1997)
4. T.B. Benjamin, P.J. Olver, Hamiltonian structure, symmetries and conservation laws for water waves. J. Fluid Mech. **125**, 137–185 (1982)
5. J.L. Bona, D. Lannes, J.-C. Saut, Asymptotic models for internal waves. J. Math. Pures Appl. **89**, 538–566 (2008)
6. A. Compelli, Hamiltonian formulation of 2 bounded immiscible media with constant non-zero vorticities and a common interface. Wave Motion **54**, 115–124 (2015)
7. A. Compelli, Hamiltonian approach to the modeling of internal geophysical waves with vorticity. Monatsh. Math. **179**(4), 509–521 (2016)
8. A. Compelli, R. Ivanov, The dynamics of flat surface internal geophysical waves with currents. J. Math. Fluid Mech. **19**(2), 329–344 (2017)
9. A. Compelli, R. Ivanov, Benjamin-Ono model of an equatorial pycnocline. Discrete Contin. Dynam. Syst. A **39**(8), 4519–4532 (2019). https://doi.org/10.3934/dcds.2019185
10. A. Constantin, J. Escher, Symmetry of steady periodic surface water waves with vorticity. J. Fluid Mech. **498**, 171–181 (2004)
11. A. Constantin, J. Escher, Analyticity of periodic traveling free surface water waves with vorticity. Ann. Math. **173**, 559–568 (2011)
12. A. Constantin, R. Ivanov, A Hamiltonian approach to wave-current interactions in two-layer fluids. Phys. Fluids **27**, 08660 (2015)
13. A. Constantin, R.S. Johnson, The dynamics of waves interacting with the Equatorial Undercurrent. Geophys. Astrophys. Fluid Dyn. **109**(4), 311–358 (2015)
14. A. Constantin, W. Strauss, Exact steady periodic water waves with vorticity. Commun. Pure Appl. Math. **57**, 481–527 (2004)
15. A. Constantin, D. Sattinger, W. Strauss, Variational formulations for steady water waves with vorticity. J. Fluid Mech. **548**, 151–163 (2006)
16. A. Constantin, R. Ivanov, E. Prodanov, Nearly-Hamiltonian structure for water waves with constant vorticity. J. Math. Fluid Mech. **9**, 1–14 (2007)
17. W. Craig, M. Groves, Hamiltonian long-wave approximations to the water-wave problem. Wave Motion **19**, 367–389 (1994)
18. W. Craig, C. Sulem, Numerical simulation of gravity waves. J. Comput. Phys. **108**, 73–83 (1993)
19. W. Craig, P. Guyenne, H. Kalisch, Hamiltonian long-wave expansions for free surfaces and interfaces. Commun. Pure Appl. Math. **58**, 1587–1641 (2005)
20. W. Craig, P. Guyenne, C. Sulem, Coupling between internal and surface waves. Nat. Hazards **57**(3), 617–642 (2011)
21. S.A. Elder, J. Williams, *Fluid Physics for Oceanographers and Physicists: An Introduction to Incompressible Flow* (Pergamon Press, Oxford, 1989)
22. A.V. Fedorov, J.N. Brown, Equatorial waves, in *Encyclopedia of Ocean Sciences* ed. by J. Steele (Academic, San Diego, 2009), pp. 3679–3695
23. A.S. Fokas, M.J. Ablowitz, The inverse scattering transform for the Benjamin-Ono equation – a pivot to multidimensional problems. Stud. Appl. Math. **68**, 1–10 (1983)
24. J.K. Hunter, B. Nachtergaele, *Applied Analysis* (World Scientific, Singapore, 2001)
25. R.S. Johnson, *Mathematical Theory of Water Waves*. Cambridge Texts in Applied Mathematics (Cambridge University Press, Cambridge, 1997)
26. R.S. Johnson, Application of the ideas and techniques of classical fluid mechanics to some problems in physical oceanography. Phil. Trans. R. Soc. A **376**, 20170092 (2018)
27. D. Milder, A note regarding "On Hamilton's principle for water waves". J. Fluid Mech. **83**, 159–161 (1977)
28. J. Miles, On Hamilton's principle for water waves. J. Fluid Mech. **83**(1), 153–158 (1977)
29. H. Ono, Algebraic solitary waves in stratified fluids. J. Phys. Soc. Jpn. **39**, 1082–1091 (1975)

30. A.F. Teles da Silva, D.H. Peregrine, Steep, steady surface waves on water of finite depth with constant vorticity. J. Fluid Mech. **195**, 281–302 (1988)
31. E. Wahlén, A Hamiltonian formulation of water waves with constant vorticity. Lett. Math. Phys. **79**, 303–315 (2007)
32. V. Zakharov, Stability of periodic waves of finite amplitude on the surface of a deep fluid (in Russian). Zh. Prikl. Mekh. Tekh. Fiz. **9**, 86–94 (1968); J. Appl. Mech. Tech. Phys. **9**, 190–194 (1968) (English translation)

Numerical Simulations of Overturned Traveling Waves

Benjamin F. Akers and Matthew Seiders

Abstract Dimension-breaking continuation as a numerical technique for computing large amplitude, overturned traveling waves is presented. Dimension-breaking bifurcations from branches of planar waves are presented in two weakly-nonlinear model equations as well as in the vortex sheet formulation of the water wave problem, with the small scale approximation (Ambrose et al., J Comput Phys 247:168–191, 2013; Akers and Reeger, Wave Motion 68:210–217, 2017). The challenges and potential of this method toward computing overturned traveling waves at the interface between three-dimensional fluids is reviewed. Numerical simulations of dimension-breaking continuation are presented in each model. Overturned traveling three-dimensional waves are presented in the vortex sheet system.

Keywords Traveling waves · Overturned · Numerical continuation

Mathematics Subject Classification (2000) 35B35, 76B15

1 Introduction

Traveling waves have a long and illustrious history with origins dating back at least to Stokes, a wonderful review of which appears in [1]. It has been known since Crapper wrote his exact solution to the capillary wave problem that there exist traveling interfacial waves in which the free surface is not a function of the horizontal coordinate [2]; we refer to these as overturned traveling waves. We will focus on the infinite depth problem in this article, but almost all of the results translate readily to finite depth, even the existence of exact solutions, for example the Kinnersley waves [3].

B. F. Akers (✉) · M. Seiders
Air Force Institute of Technology, WPAFB, Dayton, OH, USA
e-mail: benjamin.akers@afit.edu; matthew.seiders@afit.edu

© Springer Nature Switzerland AG 2019
D. Henry et al. (eds.), *Nonlinear Water Waves*, Tutorials, Schools, and Workshops in the Mathematical Sciences, https://doi.org/10.1007/978-3-030-33536-6_7

In this work we consider the mathematical and numerical difficulties associated with the computation of overturned traveling waves. This work primarily focuses on the water wave problem, but many of these same difficulties exist, without significant changes, in the hydro-elastic and internal wave cases [4–6]. We study periodic waves of the interface between two constant-density fluids undergoing irrotational motions. The fluid depth is not crucial for this discussion, but we present results only for the infinite depth case, with periodicity (of possibly very large size) in the horizontal directions. We seek traveling wave solutions, in which the free surface is of permanent form and steadily translating. The main goal being to compute waves on a two-dimensional interface, between three-dimensional fluids, which may have overhanging crests (or troughs).

Currently, no study has been conducted for fully three-dimensional water waves which are both overturned and traveling in the full equations for potential flow. A number of studies have considered overturning in the time dependent problem, for example [7–12] with a review in [13]. There are also numerous computations of permanent three-dimensional waves (both traveling and standing) in which the interface is parameterized by the horizontal coordinates, for example [14–17]. Additionally, there exist studies of axisymmetric three-dimensional overturned traveling waves in fluid jets, where such symmetry is natural [18, 19]. There have also been studies of fully three-dimensional overhanging traveling waves in model equations [20].

There are a number of reasons for the absence of previous work on three-dimensional overturned traveling waves. First, one must have a three-dimensional formulation of the problem which allows for traveling waves which are overturning. Conformal mappings are by far the most popular technique for the two-dimensional problem, but do not generalize to three-dimensions, [21–23]. There are three modern formulations which seem promising venues in which to compute overturned traveling waves in three-dimensions. These include a Hamiltonian formulation which allows for arbitrary interface parameterizations, and thus overturned interfaces, proposed by Bridges and Dias [24]. The AFM (Ablowitz, Fokas, and Musslimani) formulation has an extension for overturned interfaces [25, 26]. Finally, the vortex sheet formulation of the water wave problem can by written with an arbitrary parameterization, allowing for overturned interfaces [27–29]. More progress has been made computing overturned waves in the vortex sheet formulation than the other two. This reason for the increased simulation in the vortex sheet formulation is historic rather than strategic, and there is hope that these other formulations may be more amenable to simulation. The difficulties in numerical simulation in the vortex sheet formulation will be discussed in later sections, in a sense motivating future computational studies in alternative formulations.

A second reason for the lack of computations of overhanging three-dimensional traveling waves is the extreme expense of the computation. This cost increase comes in the natural manner of a dimension increase, but also in some more subtle ways. By far the most popular, and arguably best two-dimensional method for computing waves is via conformal mapping, which relies on complex variables, and is thus unavailable in three-dimensions. Alternatively, boundary integrals methods typically result in a Birkhoff-Rott integral, which is notoriously difficult to compute [30]. For periodic problems, the sum over periodic images can be rapidly evaluated in the two-dimensional problem, again using complex variable techniques [13]. The AFM formulation has no obvious obstacle to three-dimensional calculations, but has yet to be used for numerical computations of overturned waves; there have, however, been substantial computations of traveling non-overturned waves in this formulation [31–33]. Overturned traveling wave computations in AFM are being actively pursued [34].

In this work, we consider dimension-breaking as a continuation procedure for computing three-dimensional overturned traveling waves. The idea is to compute three-dimensional overturned traveling waves at large amplitude by first computing a large amplitude two-dimensional wave, and then slowly adding transverse variation. Naturally this first requires accurate computations of the two-dimensional profiles. One also needs to understand the manner in which transverse variation enters into these secondary bifurcations. The accuracy of two-dimensional computations of overturned waves and the asymptotics of dimension-breaking bifurcations will be both be discussed herein.

For two-dimensional fluids, with one-dimensional interfaces, a significant amount of work has been done in the study of both dynamic and steady overturned waves [35–42]. Most relevant to this work are the exact traveling solutions of Crapper [2], the numerically computed waves of Meiron and Saffman [43], and the large amplitude gravity-capillary solitary waves simulated in [42]. These represent the three qualitatively different two-dimensional overturned traveling waves from which three-dimensional overturned traveling waves can be expected to bifurcate. To date, such overhanging bifurcations have only been computed near profiles similar to those in Meiron and Saffman's work [43], which need fewer points to be resolved numerically than either the Crapper waves or the gravity-capillary solitary waves in [42].

This chapter is an outgrowth of a number of recent studies by one of the authors. For the two dimensional problem, the traveling wave ansatz developed in [28] has since been used extensively to compute two-dimensional overturning traveling waves [4–6, 29]. Examples of such waves are in Fig. 1. More recently, the three dimensional overturned traveling waves were computed in an approximate model [20]; this work uses dimension breaking to compute fully three dimensional profile whose two-dimensional cross section resembles those computed by Meiron and Saffman [43].

There has been significant recent interest in computing traveling waves in the hydro-elastic problem, where the fluid interface includes an elastic membrane [4, 5, 44–48]. The status of the field in computing overturned three-dimensional hydroelastic waves is essentially in the same state as traveling water waves. In neither setting (water waves or hydro-elastic waves) have overturned fully three-dimensional waves been computed; in both cases planar, two-dimensional, overturned waves have been computed using multiple formulations [4, 5, 28, 42, 48].

The body of the paper is organized as follows. In Sect. 2, the vortex sheet formulation is presented, including the small scale approximation. This is the

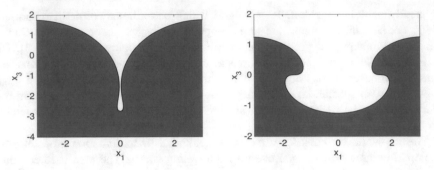

Fig. 1 Examples of extreme overturned traveling waves are depicted. The Crapper wave ($g = 0$, $At = 1$, $\tau = 2$) is on the left; a wave in the regime of Meiron and Saffman ($g = 1$, $\tau = 0$, $At = 0.1$) is on the right [43]

model for which three-dimensional overturned traveling waves are later computed. In Sect. 3, numerical dimension-breaking is presented, using the Kadomtsev–Petviashvili equation and its deep water analogue [49] for illustration purposes. In Sect. 4, three-dimensional dimension breaking continuation is presented, including a successful computation of a doubly periodic overturned traveling wave. Conclusions and future research areas are presented in Sect. 5.

2   The Vortex Sheet Formulation

In this section we present the vortex sheet formulation, in the small-scale approximation, describing the interface between two fluids which are undergoing irrotational motions (one may be a vacuum). The two fluids are separated by an interface, and are permitted to be sheared tangentially along the interface. Shear induces a jump in tangential velocity which in turn comes with an associated vorticity, whose magnitude is described by temporally and spatially varying function called the vortex sheet strength μ. We will label the interface as $X = (x_1, x_2, x_3)$. Both the interface and the vortex sheet strength will be described parametrically, as functions of α and β, and will evolve in time t. The coordinate α is aligned with the direction of propagation; β is the transverse direction. The parameterization is assumed to be isothermal with equal step length, that is

$$X_\alpha \cdot X_\beta = 0 \qquad X_\alpha^2 = X_\beta^2.$$

A more general isothermal parameterization would allow different step lengths by inserting a constant factor, typically λ, multiplying X_β^2, as in [7, 50].

The continuity equation for the interface location is

$$X_t = U\hat{\mathbf{n}} + V_1\hat{\mathbf{t}_1} + V_2\hat{\mathbf{t}_2}. \tag{2.1}$$

in which $\hat{\mathbf{n}}$ is the unit normal vector to the interface, $\hat{\mathbf{t}}_j$ are the unit tangent vectors in the α and β directions respectively. The scalars U and V_j are the interface velocity components in each of these directions. This is paired with an evolution equation for vortex sheet strength, μ, essentially a Bernoulli equation,

$$
\mu_t = \tau \kappa + \left(\frac{\mu_\alpha}{\sqrt{X_\alpha^2}} (V_1 - \mathbf{W} \cdot \hat{\mathbf{t}}_1) + \frac{\mu_\beta}{\sqrt{X_\alpha^2}} (V_2 - \mathbf{W} \cdot \hat{\mathbf{t}}_2) \right)
$$
$$
+ At \left(|\mathbf{W}|^2 + 2\mathbf{W} \cdot \hat{\mathbf{t}}_1 (V_1 - \mathbf{W} \cdot \hat{\mathbf{t}}_1) + 2\mathbf{W} \cdot \hat{\mathbf{t}}_2 (V_2 - \mathbf{W} \cdot \hat{\mathbf{t}}_2) - \frac{\mu_\alpha^2 + \mu_\beta^2}{4X_\alpha^2} - gx_3 \right)
$$

(2.2)

Here V_j are the tangential components of the velocity of the interface in the parameterized coordinates, not to be confused with $\mathbf{W} \cdot \hat{\mathbf{t}}_j$, the velocity of fluid particles on the interface. The parameter g is gravity and τ is surface tension coefficient. The parameter $At = \frac{\rho_1 - \rho_2}{\rho_1 + \rho_2}$ is the Atwood ratio, comparing the densities of the upper and lower fluids with densities of ρ_2, ρ_1, respectively; the water wave problem is the limit $At \to 1$. The vector $\mathbf{W} = (W_1, W_2, W_3)$ is the velocity of the fluid evaluated at the interface, whose closure for doubly 2π-periodic interfaces is

$$
\mathbf{W} = \frac{1}{4\pi} P.V. \sum_{n \in \mathbb{Z}} \sum_{m \in \mathbb{Z}} \int_0^{2\pi} \int_0^{2\pi} (\mu_\alpha' X'_\beta - \mu_\beta' X'_\alpha) \times \frac{(X - X' - 2\pi n \mathbf{e}_1 - 2\pi m \mathbf{e}_2)}{|X - X' - 2\pi n \mathbf{e}_1 - 2\pi m \mathbf{e}_2|^3} \, d\alpha' \, d\beta'
$$

(2.3)

in which $\mathbf{e}_j$ are the cannonical unit vectors in the jth coordinate.

For two-dimensional flows, $\mu_\beta = x_{1,\beta} = x_{3,\beta} = 0$ and $x_2 = \beta$. The sum in m and integral in β' can be evaluated exactly with elementary calculus, yielding

$$
\mathbf{W} = \frac{1}{4\pi} P.V. \sum_{n \in \mathbb{Z}} \int_0^{2\pi} \mu_\alpha' \left(\begin{pmatrix} x_3 \\ 0 \\ -x_1 \end{pmatrix} - \begin{pmatrix} x_3' \\ 0 \\ -x_1' + 2n\pi \end{pmatrix} \right) \frac{2}{(x_1 - x_1' - 2n\pi)^2 + (x_3 - x_3')^2} \, d\alpha' .
$$

(2.4)

Complexifying the domain $\mathbf{z} = x_1 + i x_3$ and using the identity,

$$
\frac{1}{2} \cot \left(\frac{\mathbf{z}}{2} \right) = \sum_{n=-\infty}^{\infty} \frac{1}{\mathbf{z} + 2n\pi},
$$

(2.5)

allows the infinite sum from (2.4) to be replaced with a hyperbolic cotangent,

$$
W_1 - i W_3 = \frac{1}{4\pi i} P.V. \int_0^{2\pi} \mu_\alpha' \cot \left(\frac{\mathbf{z} - \mathbf{z}'}{2} \right) d\alpha'
$$

(2.6)

This integral, while still singular, can be numerically approximated with standard methods, for example the alternating point trapezoid rule [51, 52]. It can also be regularized be subtraction of a Hilbert transform, see [53]. The two-dimensional problem is significantly simpler than the three-dimensional problem due fundamentally to access to the summation formulae (2.5). This formula comes from residue calculus, thus the ease of two-dimensional simulations in the vortex sheet formulation is reliant on complex variables, just like conformal mappings.

The Birkhoff-Rott integral for three-dimensional flows, (2.3), is notoriously difficult to simulate, see [7, 30]. In this work we replace it with the small-scale approximation of [7, 20],

$$\mathbf{W} \approx \frac{1}{2} H_\alpha \left[\frac{\mu_\alpha X_\beta \times X_\alpha}{\sqrt{X_\alpha^2}^3} \right] - \frac{1}{2} H_\beta \left[\frac{\mu_\beta X_\alpha \times X_\beta}{\sqrt{X_\alpha^2}^3} \right]. \tag{2.7}$$

The operators H_α and H_β are the Riesz transforms, a generalization of the Hilbert transform. The Riesz transforms have multiplicative Fourier symbols,

$$\widehat{H_\alpha f}(\mathbf{k}) = -i \frac{k_1}{\sqrt{k_1^2 + k_2^2}} \hat{f}, \quad \text{and} \quad \widehat{H_\beta f}(\mathbf{k}) = -i \frac{k_2}{\sqrt{k_1^2 + k_2^2}} \hat{f}$$

in which k_1 is the wavenumber corresponding to α and k_2 is the wavenumber corresponding to β.

When searching for traveling waves, it is convenient to parameterize in the traveling frame, so that $\mu_t = 0$. The traveling wave ansatz for waves traveling in the x_1 direction is $X_t = (c, 0, 0)$, which can be combined with (2.1), to give a closure for the interface velocities

$$U = c(\hat{\mathbf{n}})_1, \qquad V_1 = c(\hat{\mathbf{t}}_1)_1, \qquad V_2 = c(\hat{\mathbf{t}}_2)_1.$$

The prescriptions of the tangential velocities, V_j, to equal the speed times the first component of the tangent vector, $\hat{\mathbf{t}}_j$, can be thought of as being chosen to preserve the parameterization in the traveling frame. The normal velocity prescription is a restriction on the physical fluid velocity at the interface to match that of the interface itself.

Ultimately, the system of equations for computing a traveling wave in the vortex sheet formulation in three dimensions requires finding four functions x_1, x_2, x_3, and μ as well as a speed c, which solve four equations,

$$0 = \tau\kappa + \frac{1}{\sqrt{X_\alpha^2}} (\tilde{\mathbf{V}} \cdot \nabla)\mu + \text{At} \left(|\mathbf{W}|^2 + 2\mathbf{W} \cdot \hat{\mathbf{t}}_1 \tilde{V}_1 + 2\mathbf{W} \cdot \hat{\mathbf{t}}_2 \tilde{V}_2 - \frac{1}{4X_\alpha^2} |\nabla\mu|^2 - gx_3 \right),$$

$$\tag{2.8a}$$

$$0 = c(\hat{\mathbf{n}})_1 - \mathbf{W} \cdot \hat{\mathbf{n}}, \tag{2.8b}$$

$$0 = X_\alpha \cdot X_\beta, \tag{2.8c}$$

$$0 = X_\beta^2 - X_\alpha^2, \tag{2.8d}$$

in which $\tilde{V}_j = c(\hat{\mathbf{t}}_\mathbf{j})_1 - \mathbf{W} \cdot \hat{\mathbf{t}}_\mathbf{j}$, and $\tilde{\mathbf{V}} = (\tilde{V}_1, \tilde{V}_2)$ are the tangential components of the interface velocity differences. The first equation is effectively Bernoulli in the traveling frame, the second equation matches the fluid velocity to the traveling wave interface velocity, and the last two equations are descriptions of the parameterization. The system is closed by appending a scalar equation specifying a measure of the wave's amplitude. Natural amplitude measures include the crest height, total displacement, the amplitude of a Fourier mode of the third coordinate of the interface, x_3, or in dimension breaking computations the second derivative of x_3 with respect to β at the base of the trough.

3 Dimension-Breaking Continuation

From a numerical perspective, dimension-breaking is a tool to compute large amplitude traveling waves with a continuation method which pays a lower-dimensional cost for the bulk of the computation. The idea is simple and comes in three steps. First, compute large two-dimensional waves by continuation in amplitude. The trivial extension of these solutions to three dimensions yields a branch of planar waves. Second, fix an amplitude and find the transverse period from which there is a dimension-breaking bifurcation. In other words, find a wave with small transverse variation that is near to the planar wave at the prescribed amplitude. This step is the most delicate, and we will discuss it in detail in two model equations in this section. Third, continue in transverse variation, for example by using $\partial_\beta^2 x_3(0, 0)$ as a continuation parameter. The last step requires a continuation method in the higher-dimensional space, thus is numerically more expensive. It is, however, not a continuation in wave amplitude, but instead in transverse variation, thus this method may begin at a large amplitude with lower cost than continuation from small amplitude in the higher-dimensional problem directly. The search for overturned waves is done in the two-dimensional problem, thus avoiding the need to search parameter space in the more expensive three-dimensional setting.

In this section, we present numerical dimension-breaking continuation using the Kadomtsev–Petviashvili (KP) equation [54] and a model equation from [55] as pedagogical examples. These models are both derived for waves with small transverse variation, thus are a natural setting to discuss dimension-breaking. The former is a shallow water model, the latter a deep water model. Numerical computations in the KP setting are unnecessary; the entire solution set has explicit formula [56]. This problem however, can be used as a valuable test bed from which numerics can be understood, and debugged, before application to problems where exact solutions are unavailable. Neither of these pedagogical models include overturning; we will also compute dimension-breaking bifurcations in the small-scale approximation of the vortex-sheet equations in Sect. 4.

The Kadomtsev–Petviashvili (KP) equation is

$$\partial_x^2 \left(u_{xx} - cu - \frac{3}{2}u^2 \right) - u_{yy} = 0. \tag{3.1}$$

which supports solitary traveling wave solutions

$$u = -c \operatorname{sech}^2 \left(\frac{\sqrt{c}}{2}(x - ct) \right). \tag{3.2}$$

At $c = 1$, a dimension-breaking bifurcation is known explicitly,

$$u = -\frac{4(1 - \delta^2)}{4 - \delta^2} \frac{1 - \delta\cosh(ax)\cos(\omega y)}{(\cosh(ax) - \delta\cos(\omega y))^2}, \tag{3.3}$$

$$a = \sqrt{\frac{1 - \delta^2}{4 - \delta^2}}, \qquad \omega = \frac{\sqrt{3(1 - \delta^2)}}{4 - \delta^2},$$

in which the bifurcation parameter $\delta \in [0, 1)$. The transverse wavenumber, ω, is a key quantity of interest in numerical computations, as this dictates the required domain size upon which a dimension-breaking bifurcation occurs. The classic KP rescaling gives that $\omega_{\delta=0}(c) = \frac{\sqrt{3}}{4}c$.

Ignoring the above exact solution for the dimension-breaking bifurcation, the transverse wavenumber, d, and it's accompanying function, $u_1(x)$, can be calculated by linearizing the model equation using the following ansatz,

$$u = u_0(x) + \delta u_1(x) \cos(dy),$$

and linearizing with respect to δ, yields

$$\partial_x^2 \left(u_{1,xx} - cu_1 - 3u_0u_1 \right) + d^2u_1 = 0, \tag{3.4}$$

in which we seek the pair $(u_1(x), d)$.

Typically Eq. (3.4) requires numerical solution, however for KP it can be solved exactly. At $c = 1$ a solution to (3.4) is

$$u_1 = \operatorname{sech}\left(\frac{1}{2}x \right) - 2\operatorname{sech}^3\left(\frac{1}{2}x \right), \qquad \text{with,} \qquad d = \frac{\sqrt{3}}{4}.$$

This pair can be used as an initial guess for the third step, in which we continue in transverse variation. A measure of transverse variation which we have been successful using is $\partial_\beta^2 x_3(0, 0)$. An example computation is in Fig. 2.

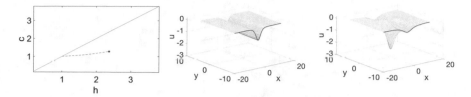

Fig. 2 Traveling waves from the Kadomtsev–Petviashvili equation along with their speed amplitude dependence. On the left, the solid (blue) curve is the planar wave speed amplitude relationship; the dashed (red) curve is the three-dimensional wave speed. In the center, is the planar wave from which the three-dimensional bifurcation occurs. On the right is the three-dimensional interface whose speed and amplitude are marked with a star in the left panel

For a general problem, one will not have access to the exact solution for either the planar wave or its bifurcation direction, (u_1, d). As an example of such an equation, we use the the deep water gravity-capillary model, from [55], presented here for waves traveling at speed c,

$$\left((2 - c)u_x + \mathcal{H}u - \mathcal{H}u_{xx} - \frac{3}{2}uu_x \right) + 2\mathcal{H}u_{yy} = 0, \quad (3.5)$$

where $\mathcal{H}$ is the Hilbert transform, whose Fourier symbol is $\widehat{\mathcal{H}(u)} = -i \, \mathrm{sign}(k)\hat{u}(k)$. The linearization about a planar solution u_0 with transverse wavenumber d gives a similar equation

$$\left((2 - c)u_{1,x} + \mathcal{H}u_1 - \mathcal{H}u_{1,xx} - \frac{3}{2}(u_0 u_{1,x} + u_1 u_{0,x}) \right) - 2d^2 \mathcal{H}u_1 = 0 \quad (3.6)$$

To solve (3.6) in this context is to compute real spectra of the operator,

$$\mathcal{L}(u_1) = (c - 2)\mathcal{H}(\partial_x u_1) - u_1 + \partial_{xx}u_1 + \frac{3}{2}\mathcal{H}(u_0\partial_x u_1 + u_{0,x}u_1) \quad (3.7)$$

We approximate $\mathcal{L}$ via Fourier-collocation, then use the QR method on the discrete approximation to $\mathcal{L}$. The computed spectrum is then searched for real eigenvalues. Any resulting pairs $(u_1(x), d)$ may then be used as an initial guess for the dimension-breaking portion of the method, wherein we continue in transverse variation (e.g. $\partial_y^2 u(0, 0)$.) An example dimension-breaking bifurcation computed in this equation is depicted in Fig. 3.

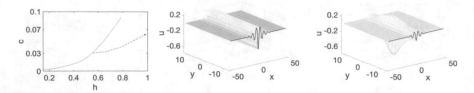

Fig. 3 Traveling waves from Eq. (3.5) along with their speed amplitude dependence. On the left, the solid (blue) curve is the planar wave speed amplitude relationship; the dashed (red) curve is the three-dimensional wave speed. In the center, is the planar wave from which the three-dimensional bifurcation occurs. On the right is the three-dimensional interface whose speed and amplitude are marked with a star in the left panel

4 Overturning Three-Dimensional Traveling Waves

In this section we present three-dimensional computations of overturned waves in the vortex sheet equations. As discussed earlier, the Birkhoff-Rott integral is very difficult to simulate for three-dimensional periodic waves, see [30]. There have not yet been successful computations of fully (non-planar) three-dimensional periodic overturned traveling waves; in this section we present computations using the small-scale approximation of [7].

Three-dimensional traveling waves in system (2.8) have been computed using the small-scale approximation (2.7), resulting in similar waves and method to those presented in [20]. The continuation method applied here uses a dimension-breaking approach, but absent knowledge of the linearization about a planar wave as discussed in the previous section. Such knowledge would lead to a more systematic and reliable continuation procedure, but as of yet this approach has not been implemented. Instead, a single dimension-breaking bifurcation is found via trial and error, and then numerical continuation is used to find nearby dimension-breaking bifurcations with similar amplitudes and periods.

A single three-dimensional bifurcation is presented here as an illustrative example. In this example, overturned traveling waves exist on the interface between a small density fluid and a higher density fluid, as first simulated by Meiron and Saffman [43]. The profiles in this regime are quite regular, and need relatively few points for accurate simulation (as compared to the profile in Fig. 3). This regime is thus a good candidate for three dimensional simulations.

In Fig. 4, the pure gravity waves with near to equal mass fluids in the regime first computed by Meiron and Saffman [43] are computed. The planar waves in this regime overturn at large amplitude, and are very regular. Dimension-breaking bifurcations were computed at a sampling of amplitudes, and the transverse period's amplitude dependence is reported. In this computation, a dimension-breaking bifurcation was guessed via trial and error near $h \approx 0.1$. There is potential for

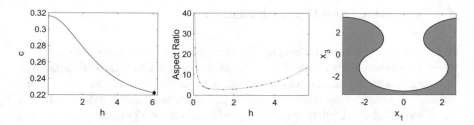

Fig. 4 Traveling waves with $At = 0.1$, $g = 1$, $\tau = 0$. The overturning waves in this parameter regime were discovered by Meiron and Saffman. The speed amplitude curve of planar traveling waves is in the left panel. The aspect ratio (transverse/longitudinal period) of computed dimension-breaking bifurcations is in the center panel. An example of an overturning planar profile, marked with a diamond in the left panel, is in the right panel

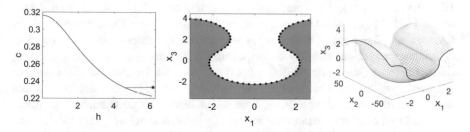

Fig. 5 Periodic traveling solution to system (2.8) with $At = 0.1$, $g = 1$, $\tau = 2$ are depicted. On the left, the solid (blue) curve is the planar wave speed amplitude relationship; the dashed (red) curve is the speed of the three-dimensional waves. A three dimensional wave marked with the star in the left panel is sliced at at its centerline, $x_2 = 0$, in the center panel, with the full wave in the right panel

the initial guess to be avoided via small amplitude arguments, as in [57], however we fundamentally desire direct access to the transverse period at large amplitude.

Figure 4, illustrates the need for the operator-spectrum based approach outlined in the previous section. First, following the transverse period from small amplitude is numerically costly; continuation for the transverse period in system (2.8) requires use of a three-dimensional solver with a number of simulations proportionate to the number of steps in amplitude. Even worse, there is no guarantee that continuation from small amplitude produces a large amplitude overturned wave.

An example of a parameter set where such brute force guessing has been successful in computing overturned traveling waves is in Fig. 5. The overturned waves in this regime are qualitatively similar to those computed by Meiron and Saffman, however their bifurcation structure is simpler. The transverse period is continuous as a function of amplitude, and large amplitude dimension-breaking bifurcations were found.

5   Conclusions and Future Research

In this work, we present techniques for computing three-dimensional overturned traveling waves on fluid interfaces. The prospects and limitations of dimension-breaking as a numerical continuation technique is evaluated with this aim in mind. Numerical computations were presented for two-dimensional interfaces in the full water wave problem. Dimension-breaking bifurcations were presented in the weakly nonlinear models [54, 55] and in the vortex sheet formulation with the small-scale approximation to the Birkhoff-Rott integral [7]. Dimension-breaking is a promising continuation technique, whose importance grows with cost of three-dimensional computation, thus should be of particular use for computing three-dimensional overturned traveling waves in the Euler equations.

Disclaimer This report was prepared as an account of work sponsored by an agency of the United States Government. Neither the United States Government nor any agency thereof, nor any of their employees, make any warranty, express or implied, or assumes any legal liability or responsibility for the accuracy, completeness, or usefulness of any information, apparatus, product, or process disclosed, or represents that its use would not infringe privately owned rights. Reference herein to any specific commercial product, process, or service by trade name, trademark, manufacturer, or otherwise does not necessarily constitute or imply its endorsement, recommendation, or favoring by the United States Government or any agency thereof. The views and opinions of authors expressed herein do not necessarily state or reflect those of the United States Government or any agency thereof.

Acknowledgements Benjamin Akers and Matthew Seiders were supported in part by the Air Force Office of Scientific Research (AFOSR) and the Office of Naval Research (ONR) during the preparation of this manuscript.

References

1. A.D.D. Craik, The origins of water wave theory. Annu. Rev. Fluid Mech. **36**, 1–28 (2004)
2. G.D. Crapper, An exact solution for progressive capillary waves of arbitrary amplitude. J. Fluid Mech. **2**, 532–540 (1957)
3. W. Kinnersley, Exact large amplitude capillary waves on sheets of fluid. J. Fluid Mech. **77**(2), 229–241 (1976)
4. B.F. Akers, D.M. Ambrose, D.W. Sulon, Periodic traveling interfacial hydroelastic waves with or without mass. Zeitschrift für angewandte Mathematik und Physik **68**(6), 141 (2017)
5. B.F. Akers, D.M. Ambrose, D.W. Sulon, Periodic travelling interfacial hydroelastic waves with or without mass II: multiple bifurcations and ripples. Eur. J. Appl. Math. **30**, 1–35 (2018)
6. B.F. Akers, D.M. Ambrose, K. Pond, J.D. Wright, Overturned internal capillary-gravity waves. Eur. J. Mech. B. Fluids **57**, 143–151 (2016)
7. D.M. Ambrose, M. Siegel, S. Tlupova, A small-scale decomposition for 3D boundary integral computations with surface tension. J. Comput. Phys. **247**, 168–191 (2013)
8. S.T. Grilli, P. Guyenne, F. Dias, A fully non-linear model for three-dimensional overturning waves over an arbitrary bottom. Int. J. Numer. Methods Fluids **35**(7), 829–867 (2001)

9. P. Lubin, S. Vincent, S. Abadie, J.-P. Caltagirone, Three-dimensional large eddy simulation of air entrainment under plunging breaking waves. Coast. Eng. **53**(8), 631–655 (2006)
10. M. Xue, H. Xü, Y. Liu, D.K.P. Yue, Computations of fully nonlinear three-dimensional wave–wave and wave–body interactions. Part 1. dynamics of steep three-dimensional waves. J. Fluid Mech. **438**, 11–39 (2001)
11. C. Fochesato, F. Dias, A fast method for nonlinear three-dimensional free-surface waves, in *Proceedings of the Royal Society of London A: Mathematical, Physical and Engineering Sciences*, vol 462 (The Royal Society, London, 2006), pp. 2715–2735
12. T.Y. Hou, P. Zhang, Convergence of a boundary integral method for 3-D water waves. Discrete Contin. Dynam. Systems Series B **2**(1), 1–34 (2002)
13. S.T. Grilli, F. Dias, P. Guyenne, C. Fochesato, F. Enet, Progress in fully nonlinear potential flow modeling of 3D extreme ocean waves, in *Advances in Numerical Simulation of Nonlinear Water Waves* (World Scientific, Singapore, 2010), pp. 75–128
14. D.I. Meiron, P.G. Saffman, H.C. Yuen, Calculation of steady three-dimensional deep-water waves. J. Fluid Mech. **124**, 109–121 (1982)
15. C.H. Rycroft, J. Wilkening, Computation of three-dimensional standing water waves. J. Comput. Phys. **255**, 612–638 (2013)
16. D.P. Nicholls, F. Reitich, Stable, high-order computation of traveling water waves in three dimensions. Eur. J. Mech. B. Fluids **25**(4), 406–424, 2006
17. E.I. Parau, J.-M. Vanden-Broeck, M.J. Cooker, Nonlinear three-dimensional gravity–capillary solitary waves. J. Fluid Mech. **536**, 99–105 (2005)
18. J-M Vanden-Broeck, T. Miloh, B. Spivack, Axisymmetric capillary waves. Wave Motion **27**(3), 245–256 (1998)
19. S. Grandison, J.-M. Vanden-Broeck, D.T. Papageorgiou, T. Miloh, B. Spivack, Axisymmetric waves in electrohydrodynamic flows. J. Eng. Math. **62**(2), 133–148 (2008)
20. B.F. Akers, J.A. Reeger, Three-dimensional overturned traveling water waves. Wave Motion **68**, 210–217 (2017)
21. A.I. Dyachenko, E.A. Kuznetsov, M.D. Spector, V.E. Zakharov, Analytical description of the free surface dynamics of an ideal fluid (canonical formalism and conformal mapping). Phys. Lett. A **221**(1–2), 73–79 (1996)
22. S.A. Dyachenko, On the dynamics of a free surface of an ideal fluid in a bounded domain in the presence of surface tension. J. Fluid Mech. **860**, 408–418 (2019)
23. T. Gao, P. Milewski, J.-M. Vanden-Broeck, Hydroelastic solitary waves with constant vorticity. Wave Motion **85**, 84–97 (2018)
24. F. Dias, T.J. Bridges, The numerical computation of freely propagating time-dependent irrotational water waves. Fluid Dyn. Res. **38**(12), 803–830 (2006)
25. M.J. Ablowitz, A.S. Fokas, Z.H. Musslimani, On a new non-local formulation of water waves. J. Fluid Mech. **562**, 313–343 (2006)
26. A.C.L. Ashton, A.S. Fokas, A non-local formulation of rotational water waves. J. Fluid Mech. **689**, 129–148 (2011)
27. D.M. Ambrose, N. Masmoudi, et al., Well-posedness of 3D vortex sheets with surface tension. Commun. Math. Sci. **5**(2), 391–430 (2007)
28. B. Akers, D.M. Ambrose, J.D. Wright, Traveling waves from the arclength parameterization: Vortex sheets with surface tension. Interfaces Free. Bound. **15**, 359–380 (2013)
29. B.F. Akers, D.M. Ambrose, J.D. Wright, Gravity perturbed crapper waves. Proc. R. Soc. London, Ser. A **470**(2161), 20130526 (2014)
30. J. Beale, A convergent boundary integral method for three-dimensional water waves. Math. Comput. **70**(235), 977–1029 (2001)
31. B. Deconinck, K. Oliveras, The instability of periodic surface gravity waves. J. Fluid Mech. **675**, 141–167 (2011)
32. K. Oliveras, B. Deconinck, The instabilities of periodic traveling water waves with respect to transverse perturbations. Nonlinear Wave Equ. **635**, 131 (2015)
33. B. Deconinck, O. Trichtchenko, Stability of periodic gravity waves in the presence of surface tension. Eur. J. Mech. B. Fluids **46**, 97–108 (2014)

34. K. Oliveras, Personal communication
35. J.Y. Holyer, Large amplitude progressive interfacial waves. J. Fluid Mech. **93**(3), 433–448 (1979)
36. REL Turner, J.-M. Vanden-Broeck, The limiting configuration of interfacial gravity waves. Phys. Fluids **29**(2), 372–375 (1986)
37. S. Koshizuka, A. Nobe, Y. Oka, Numerical analysis of breaking waves using the moving particle semi-implicit method. Int. J. Numer. Methods Fluids **26**(7), 751–769 (1998)
38. O.B. Fringer, R.L. Street, The dynamics of breaking progressive interfacial waves. J. Fluid Mech. **494**, 319–353 (2003)
39. G. Chen, C. Kharif, S. Zaleski, J. Li, Two-dimensional navier–stokes simulation of breaking waves. Phys. Fluids **11**(1), 121–133 (1999)
40. S.T. Grilli, P. Guyenne, F. Dias, A fully non-linear model for three-dimensional overturning waves over an arbitrary bottom. Int. J. Numer. Methods Fluids **35**(7), 829–867 (2001)
41. P. Guyenne, S.T. Grilli, Numerical study of three-dimensional overturning waves in shallow water. J. Fluid Mech. **547**, 361–388 (2006)
42. Z. Wang, Stability and dynamics of two-dimensional fully nonlinear gravity–capillary solitary waves in deep water. J. Fluid Mech. **809**, 530–552 (2016)
43. D.I. Meiron, P.G. Saffman, Overhanging interfacial gravity waves of large amplitude. J. Fluid Mech. **129**, 213–218 (1983)
44. T. Gao, J.-M. Vanden-Broeck, Z. Wang, Numerical computations of two-dimensional flexural-gravity solitary waves on water of arbitrary depth. IMA J. Appl. Math. **83**(3), 436–450 (2018)
45. T. Gao, J.-M. Vanden-Broeck, Numerical studies of two-dimensional hydroelastic periodic and generalised solitary waves. Phys. Fluids **26**(8), 087101 (2014)
46. P.A. Milewski, J.-M. Vanden-Broeck, Z. Wang, Hydroelastic solitary waves in deep water. J. Fluid Mech. **679**, 628–640 (2011)
47. P. Guyenne, E. Parau, Forced and unforced flexural-gravity solitary waves. Procedia IUTAM **11**, 44–57 (2014)
48. T. Gao, Z. Wang, J.-M. Vanden-Broeck, New hydroelastic solitary waves in deep water and their dynamics. J. Fluid Mech. **788**, 469–491 (2016)
49. B. Akers, P.A. Milewski, A model equation for wavepacket solitary waves arising from capillary-gravity flows. Stud. Appl. Math. **122**(3), 249–274 (2009)
50. E. Aulisa, M. Toda, Z.S. Kose, Constructing isothermal curvature line coordinates on surfaces which admit them. Cent. Eur. J. Math. **11**(11), 1982–1993 (2013)
51. J.T. Beale, T.Y. Hou, J. Lowengrub, Convergence of a boundary integral method for water waves. SIAM J. Numer. Anal. **33**(5), 1797–1843 (1996)
52. L.N. Trefethen, J.A.C. Weideman, The exponentially convergent trapezoidal rule. SIAM Rev. **56**(3), 385–458 (2014)
53. D.M. Ambrose, J. Wilkening, Computation of symmetric, time-periodic solutions of the vortex sheet with surface tension. Proc. Natl. Acad. Sci. **107**(8), 3361–3366 (2010)
54. B.B. Kadomtsev, V.I. Petviashvili, On the stability of solitary waves in weakly dispersing media. Sov. Phys. Dokl. **15**, 539–541 (1970)
55. B. Akers, P.A. Milewski, A model equation for wavepacket solitary waves arising from capillary-gravity flows. Stud. Appl. Math. **122**(3), 249–274 (2009)
56. M.D. Groves, M. Haragus, S.M. Sun, A dimension–breaking phenomenon in the theory of steady gravity–capillary water waves. Philos. Trans. R. Soc. Lond. A: Math. Phys. Eng. Sci. **360**(1799), 2189–2243 (2002)
57. P.A. Milewski, Z. Wang, Transversally periodic solitary gravity–capillary waves. Proc. R. Soc. A Math. Phys. Eng. Sci. **470**, 20130537 (2014)

A Model for the Periodic Water Wave Problem and Its Long Wave Amplitude Equations

Roman Bauer, Patrick Cummings, and Guido Schneider

Abstract We are interested in the validity of the KdV and of the long wave NLS approximation for the water wave problem over a periodic bottom. Approximation estimates are non-trivial, since solutions of order $\mathcal{O}(\varepsilon^2)$, resp. $\mathcal{O}(\varepsilon)$, have to be controlled on an $\mathcal{O}(1/\varepsilon^3)$, resp. $\mathcal{O}(1/\varepsilon^2)$, time scale. In contrast to the spatially homogeneous case, in the periodic case new quadratic resonances occur and make a more involved analysis necessary. For a phenomenological model we present some results and explain the underlying ideas. The focus is on results which are robust in the sense that they hold under very weak non-resonance conditions without a detailed discussion of the resonances. This robustness is achieved by working in spaces of analytic functions. We explain that, if analyticity is dropped, the KdV and the long wave NLS approximation make wrong predictions in case of unstable resonances and suitably chosen periodic boundary conditions. Finally we outline, how, we think, the presented ideas can be transferred to the water wave problem.

Keywords KdV approximation · NLS approximation · Error estimates

Mathematics Subject Classification (2000) Primary 76B15; Secondary 35Q53, 35Q55

R. Bauer · G. Schneider (✉)
Institut für Analysis, Dynamik und Modellierung, Universität Stuttgart, Stuttgart, Germany
e-mail: guido.schneider@mathematik.uni-stuttgart.de

P. Cummings
Department of Mathematics and Statistics, Boston University, Boston, MA, USA

© Springer Nature Switzerland AG 2019
D. Henry et al. (eds.), *Nonlinear Water Waves*, Tutorials, Schools, and Workshops in the Mathematical Sciences, https://doi.org/10.1007/978-3-030-33536-6_8

1 Introduction

The two-dimensional irrotational water wave problem with a free surface $\Gamma(t) = \{y = \eta(x, t) : x \in \mathbb{R}\}$ over an L-periodic bottom $\mathcal{B} = \{y = b(x) : b(x) = b(x + L), x \in \mathbb{R}\}$ is governed by a system of nonlinear PDEs which are given by

$$\partial_x^2 \phi + \partial_y^2 \phi = 0, \quad \text{in } \Omega(t),$$

$$\partial_{\vec{n}} \phi = 0, \quad \text{on } \mathcal{B},$$

$$\partial_t \eta = \partial_y \phi - (\partial_x \eta)\partial_x \phi, \quad \text{on } \Gamma(t),$$

$$\partial_t \phi = -\frac{1}{2}((\partial_x \phi)^2 + (\partial_y \phi)^2) + \mu \partial_x \left(\frac{\partial_x \eta}{\sqrt{1 + (\partial_x \eta)^2}} \right) - g\eta, \quad \text{on } \Gamma(t),$$

for the flow potential ϕ and the elevation of the top surface η, where $\Omega(t) = \{(x, y) : b(x) < y < \eta(x, t)\}$, where g is the gravitational acceleration, and where $\mu \geq 0$ is the surface tension parameter. For a non-dimensionalized version see [3]. It is well known that the water wave problem is completely described by the elevation η of the top surface and the horizontal velocity $w = \partial_x \phi|_\Gamma$ at the top surface $\Gamma(t)$ (Fig. 1).

We are interested in the qualitative behavior of the solutions:

- The linearized problem is solved by Bloch modes, cf. [8],

$$\binom{\eta}{w} = e^{ilx} f_n(l, x) e^{i\omega_n(l)t},$$

with $n \in \mathbb{Z} \setminus \{0\}$, $f_n(l, x) = f_n(l, x + L) \in \mathbb{C}^2$, and $l \in \left[-\frac{\pi}{L}, \frac{\pi}{L}\right]$. Curves of eigenvalues $\omega_n(l)$ are sketched in Fig. 2. They are ordered as $\omega_n(l) \leq \omega_{n+1}(l)$ with $\omega_{-n}(l) = -\omega_n(l)$ for $n \in \mathbb{Z} \setminus \{-1, 0, 1\}$ and $\omega_1(l) = \omega_{-1}(-l)$. Due to the periodicity of the bottom, spectral gaps can occur.
- With the ansatz

$$\binom{\eta}{w} = \varepsilon^2 A(\varepsilon(x - ct), \varepsilon^3 t) f_1(0, x) \tag{1.1}$$

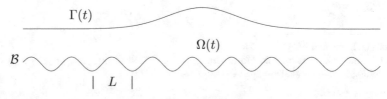

Fig. 1 The water wave problem over an L-periodic bottom

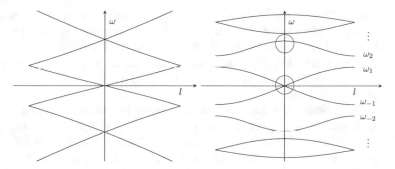

Fig. 2 The panels show the curves of eigenvalues $l \mapsto \omega_n(l)$, $n \in \mathbb{Z} \setminus \{0\}$ of the linearized water wave problem. The left panel shows the curves of eigenvalues in the homogeneous case, i.e., $b(x) = $ const., in case of positive surface tension. For $L = 2\pi$ the dispersion relation $\omega^2 = (k + \mu k^3) \tanh(k)$ in Fourier space transfers to Bloch space by setting $k = n + l$ with $n \in \mathbb{Z}$. In case of a periodic bottom, cf. [8], spectral gaps, such as sketched in the right panel, can occur. The modes in the blue circle can be described by a KdV approximation. The modes in the red circle can be described by an NLS approximation. For the derivation of the NLS equation it is essential that $\omega_2(0) > 0$

a KdV equation

$$\partial_T A = \nu_1 \partial_X^3 A + \nu_2 \partial_X (A^2), \qquad (1.2)$$

can be derived, with amplitude $A(X, T) \in \mathbb{R}$, with group velocity $c \in \mathbb{R}$, with $0 < \varepsilon \ll 1$ a small perturbation parameter, and with coefficients $\nu_1, \nu_2 \in \mathbb{R}$.

• With the ansatz

$$\begin{pmatrix} \eta \\ w \end{pmatrix} = \varepsilon A(\varepsilon x, \varepsilon^2 t) f_n(0, x) e^{i \omega_n(0) t} + \text{c.c.} \qquad (n \neq \pm 1) \qquad (1.3)$$

an NLS equation

$$i \partial_T A = \nu_1 \partial_X^2 A + \nu_2 A |A|^2, \qquad (1.4)$$

can be derived, with amplitude $A(X, T) \in \mathbb{C}$, with $0 < \varepsilon \ll 1$ a small perturbation parameter, and with coefficients $\nu_1, \nu_2 \in \mathbb{R}$.

Our future goal is to prove error estimates between these approximations and true solutions of the water wave problem. Such estimates are a nontrivial task since for the KdV approximation we have to control solutions of order $\mathcal{O}(\varepsilon^2)$ on an $\mathcal{O}(1/\varepsilon^3)$-time scale, and for the NLS approximation we have to control solutions of order $\mathcal{O}(\varepsilon)$ on an $\mathcal{O}(1/\varepsilon^2)$-time scale.

(A) In the homogeneous case, $b(x) = -1$, there are two fundamentally different approaches to prove KdV approximation results. For solutions to the KdV equation with analytic initial conditions a Cauchy–Kowalevskaya based approach can be chosen, see [15, 18]. Working in spaces of analytic functions gives some artificial smoothing which allows to gain the above explained missing order with respect to ε via the derivative in front of the nonlinear terms in the KdV equation. This 'analytic' approach is very robust and works without a detailed analysis of the underlying problem, but doesn't give optimal results.

For initial conditions in Sobolev spaces the underlying idea to gain such estimates is conceptually rather simple, namely the construction of a suitably chosen energy which includes the terms of order $\mathcal{O}(\varepsilon^2)$ in the equation for the error, such that for the energy finally $\mathcal{O}(\varepsilon^3 t)$ growth rates occur. However, the method is less robust since for every single original system a different energy occurs and the major difficulty is the construction of this energy. Estimates that the formal KdV approximation and true solutions of the different formulations of the homogeneous water wave problem stay close together over the natural KdV time scale have been shown for instance in [7, 9, 22, 23] by using this approach.

(B) In the homogeneous case, $b(x) = -1$, the NLS approximation has been justified in various papers for a number of original systems, cf. [14, 16, 19]. If no quadratic terms are present in the original system a simple application of Gronwall's inequality allows to prove the validity of the NLS approximation. Quadratic terms can be eliminated by a near identity change of variables, if a non-resonance condition is satisfied. This non resonance condition has been weakened in a number of papers, cf. [20]. The NLS approximation has been justified for the two-dimensional irrotational water wave problem in case of infinite depth and no surface tension [25, 26], and in case of finite depth and no surface tension [11].

(A+B) KdV approximation results in the spatially periodic case are only known for small perturbations of a flat bottom, cf. [3, 4, 13]. To our knowledge NLS approximation results in the spatially periodic case do not exist for the water wave problem.

It is the purpose of this paper to present for a phenomenological model, which has similar properties as the water wave problem, some approximation results and the underlying ideas of their proofs. One focus is on results which are robust in the sense that they hold under very weak non-resonance conditions without a detailed discussion of the resonances. This robustness is achieved by working in spaces of analytic functions. We explain that, if analyticity is dropped, the KdV approximation and the long wave NLS approximation make wrong predictions in case of unstable resonances and suitably chosen periodic boundary conditions. Finally we outline how, we think, the presented ideas can be transferred to the water wave problem.

2 The Boussinesq Klein–Gordon Model

In Bloch space the two-dimensional irrotational water wave problem is formally of the form

$$\partial_t \tilde{u}_n(l, t) = i\omega_n(l)\tilde{u}_n(l, t)$$

$$\text{ă} \quad +i \sum_{n_1, n_2 \in \mathbb{Z}\backslash\{0\}} \int_{-\frac{\pi}{L}}^{\frac{\pi}{L}} \beta_{n,n_1,n_2}(l, l - l_1, l_1)\tilde{u}_{n_1}(l - l_1, t)\tilde{u}_{n_2}(l_1, t)dl_1 + \ldots,$$

with $n \in \mathbb{Z}\backslash\{0\}$, $\tilde{u}_n(l,) = \tilde{u}_n(l + \frac{2\pi}{L}, t)$, and nonlinear kernels $\beta_{n,n_1,n_2}(l, l-l_1, l_1) \in \mathbb{R}$, cf. [21]. The model, which we consider, is a Boussinesq equation coupled with a Klein-Gordon equation, in the following called BKG system. It possesses a Fourier mode representation which shares several properties with the above Bloch wave representation of the water wave problem. It is given by

$$\partial_t^2 u = \alpha^2 \partial_x^2 u + \partial_t^2 \partial_x^2 u + \partial_x^2(a_{uu}u^2 + 2a_{uv}uv + a_{vv}v^2), \qquad (2.1)$$

$$\partial_t^2 v = \partial_x^2 v - v + b_{uu}u^2 + 2b_{uv}uv + b_{vv}v^2, \qquad (2.2)$$

with $u = u(x, t)$, $v = v(x, t)$, $x, t \in \mathbb{R}$, and coefficients $\alpha > 0$, $a_{uu}, \ldots, b_{vv} \in \mathbb{R}$. The curves of eigenvalues are given by

$$\omega_u(k) = \frac{\alpha k}{\sqrt{1 + k^2}} \quad \text{and} \quad \omega_v(k) = \sqrt{1 + k^2}. \qquad (2.3)$$

Hence, the spectral picture of the water wave problem over a periodic bottom, which is qualitatively sketched in the right panel of Fig. 2, and of the BKG system, see Fig. 3, look qualitatively the same. Moreover, in both systems the nonlinear terms vanish for modes associated to ω_{+1}, resp. ω_u, at the wave numbers $l = 0$, resp. $k = 0$. Since the subsequent non-resonance conditions (3.1) and (3.5) come from the Bloch/Fourier mode representations of the original systems they are the same for all original systems with a spectral picture as plotted in Fig. 2.

Inserting the ansatz

$$\varepsilon^2 \psi_u^{KdV}(x, t) = \varepsilon^2 A(\varepsilon(x - \alpha t), \varepsilon^3 t) \quad \text{and} \quad \varepsilon^2 \psi_v^{KdV} = 0 \qquad (2.4)$$

into (2.1)–(2.2) yields the KdV equation

$$\partial_T A = \nu_1 \partial_X^3 A + \nu_2 \partial_X(A^2), \qquad (2.5)$$

with coefficients $\nu_1, \nu_2 \in \mathbb{R}$, the slow temporal variable $T = \varepsilon^3 t$, and the long spatial variable $X = \varepsilon(x - \alpha t)$.

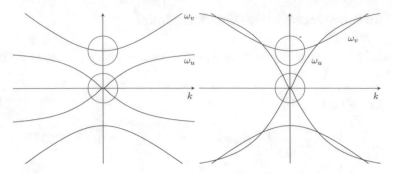

Fig. 3 The curves of eigenvalues $\pm\omega_u$, $\pm\omega_v$ for the linearized BKG system plotted as a function over the Fourier wave numbers in case $\alpha^2 = 1$ (left) and $\alpha^2 = 5$ (right). The modes in the blue circles are described by the KdV approximation. The modes in the red circles are described by the NLS approximation

Inserting the ansatz

$$\varepsilon\psi_u^{\text{NLS}}(x,t) = \mathcal{O}(\varepsilon^2) \qquad \text{and} \qquad \varepsilon\psi_v^{\text{NLS}} = \varepsilon A(\varepsilon x, \varepsilon^2 t)e^{it} + c.c. + \mathcal{O}(\varepsilon^2)$$
(2.6)

into (2.1)–(2.2) yields the NLS equation

$$i\partial_T A = \nu_1 \partial_X^2 A + |A|^2 A = 0,$$
(2.7)

with coefficients $\nu_1, \nu_2 \in \mathbb{R}$, the slow temporal variable $T = \varepsilon^2 t$, and the long spatial variable $X = \varepsilon x$. The ansatz is called long wave NLS approximation since we have $k_0 = 0$ for the wave number of the underlying carrier wave $e^{i(k_0 x + \omega_0 t)}$.

We are interested in the validity of the KdV approximation and long wave NLS approximation for the BKG system. For this phenomenological model we present some approximation results and explain the underlying ideas. Approximation estimates are non-trivial, since solutions of order $\mathcal{O}(\varepsilon^2)$, resp. $\mathcal{O}(\varepsilon)$, have to be controlled on an $\mathcal{O}(1/\varepsilon^3)$, resp. $\mathcal{O}(1/\varepsilon^2)$, time scale. That these approximation results are really subtle is explained in the next section when the resonance structure of the problem is discussed.

3   The Resonance Structure

The BKG system is written as first order system

$$\partial_t U = \Lambda U + N(U, U),$$

where ΛU stands for the linear terms and where the nonlinear terms are represented by the symmetric bilinear mapping $N(U, U)$.

3.1 Resonances in the KdV Case

The error $\varepsilon^\beta R = U - \varepsilon^2 \psi^{\text{KdV}}$ made by the KdV approximation $\varepsilon^2 \psi^{\text{KdV}}$ satisfies

$$\partial_t R = \Lambda R + 2\varepsilon^2 N(\psi^{\text{KdV}}, R) + \mathcal{O}(\varepsilon^3),$$

where $\mathcal{O}(\varepsilon^3)$ contains the nonlinear terms with respect to R and the residual terms, i.e., the terms which do not cancel after inserting the KdV approximation into the BKG system. In order to obtain $\mathcal{O}(\varepsilon^{3+\beta})$ for the residual terms in this equation, higher order terms have to be added to the KdV approximation $\varepsilon^2 \psi^{\text{KdV}}$. This is standard and so we will concentrate on other aspects. Due to the term $2\varepsilon^2 N(\psi^{\text{KdV}}, R)$ a simple application of Gronwall's inequality is not sufficient to obtain an $\mathcal{O}(1)$-bound for R on the long $\mathcal{O}(1/\varepsilon^3)$-time scale. The difficulty can be overcome by normal form transformations and energy estimates. In this section we will concentrate on the normal form transformations, i.e., near identity change of variables, and on the resonances which prevent the elimination of the quadratic terms by normal form transformations. A term $\psi^{\text{KdV}} R_j$ can be eliminated in the i-th equation with a near identity change of variables

$$R_i = \widetilde{R}_i + \varepsilon^2 M(\psi^{\text{KdV}}, \widetilde{R}_j),$$

with M a suitably chosen bilinear mapping, if the non-resonance condition

$$\omega_i(k) \neq \omega_j(k) \tag{3.1}$$

is satisfied for all $k \in \mathbb{R}$. Herein, ω_j is the curve of eigenvalues corresponding to R_j. Hence, in the R_u-equation the term $2\psi^{\text{KdV}} R_u$ cannot be eliminated. If only this term is resonant, it can be controlled with energy estimates. However, for a coefficient $\alpha > 2$ in (2.1) there are $k_1, k_2 > 0$ with $\omega_u(k_j) = \omega_v(k_j)$ for $j = 1, 2$, see the right panel of Fig. 3. Hence, the terms $2\psi^{\text{KdV}}(0)R_v(k_j)$ for $j = 1, 2$ cannot be eliminated in the R_u-equation.

Similarly, in the R_v-equation the term $2\psi^{\text{KdV}} R_v$ cannot be eliminated. If only this term is resonant, it can be controlled with energy estimates. The fact, that $\omega_u(k_j) = \omega_v(k_j)$ for $j = 1, 2$, implies now also that the terms $2\psi^{\text{KdV}}(0)R_u(k_j)$ for $j = 1, 2$ cannot be eliminated in the R_v-equation.

These resonances can be used to prove that in case of $2\pi/k_1$-periodicity, with $k_2 \notin k_1\mathbb{N}$, the KdV equation makes wrong predictions about the dynamics of the

BKG system. In order to illustrate this, we make the ansatz

$$u = \varepsilon^2 A(\varepsilon^2 t) + \varepsilon^n A_1(\varepsilon^2 t)e^{i\omega_u(k_1)t}e^{ik_1 x} + \varepsilon^n A_{-1}(\varepsilon^2 t)e^{-i\omega_u(-k_1)t}e^{-ik_1 x},$$

$$v = \varepsilon^n B_1(\varepsilon^2 t)e^{i\omega_v(k_1)t}e^{ik_1 x} + \varepsilon^n B_{-1}(\varepsilon^2 t)e^{i\omega_v(-k_1)t}e^{-ik_1 x},$$

to analyze the resonance at the wave number $k = k_1$. Equating the coefficients to zero at $\varepsilon^4 e^{0it}e^{0ix}$ in the u-equation and at $\varepsilon^{n+2}e^{i\omega_u(k_1)t}e^{ik_1 x}$ both in the u- and v-equation yields, with $\tau = \varepsilon^2 t$, that

$$\partial_\tau^2 A = 0, \tag{3.2}$$

$$2i\omega_u(k_1)\partial_\tau A_1 = -2k_1^2(a_{uu}AA_1 + a_{uv}AB_1), \tag{3.3}$$

$$2i\omega_v(k_1)\partial_\tau B_1 = 2(b_{uu}AA_1 + b_{uv}AB_1). \tag{3.4}$$

The first equation is the KdV equation, i.e., (2.5) restricted to the wave number $k = 0$. Hence, for instance on a $\mathcal{O}(\varepsilon^{-1/2})$-time scale with respect to τ, the variable A can be considered to be constant in time. The last two equations can be written as

$$\partial_\tau \begin{pmatrix} A_1 \\ B_1 \end{pmatrix} = M\breve{a} \begin{pmatrix} A_1 \\ B_1 \end{pmatrix},$$

with

$$M = \frac{1}{i\omega_u(k_1)} \begin{pmatrix} -a_{uu}k_1^2 A & -a_{uv}k_1^2 A \\ b_{uu}A & b_{uv}A \end{pmatrix}.$$

By choosing the real-valued coefficients a_{uu}, a_{uv}, b_{uu}, and b_{uv} in a suitable way, namely

$$(a_{uu}k_1^2 + b_{uv})^2 - a_{uv}b_{uu}k_1^2 < 0,$$

the matrix M has an eigenvalue with strictly positive real part. Therefore, the occurrence of such an eigenvalue is excluded in case $a_{uv}b_{uu} \leq 0$. Hence, for suitably chosen coefficients growth rates $e^{\beta\tau} = e^{\beta\varepsilon^2 t} = e^{\beta T/\varepsilon}$ with a $\beta > 0$ occur. These allow us to bring $\varepsilon^n A_1$ and $\varepsilon^n B_1$, which are initially of order $\mathcal{O}(\varepsilon^n)$, to an order $\mathcal{O}(\varepsilon^2)$ at a time $T = \mathcal{O}((n-2)\varepsilon|\ln(\varepsilon)|) \ll 1$. Therefore, we have that $v = \mathcal{O}(\varepsilon^2)$ far before the natural time scale of the KdV equation. Hence, in this situation the KdV approximation makes wrong predictions. These calculations can be transferred into a rigorous proof of a non-approximation result in case of $\frac{2\pi}{k_1}$-periodic boundary conditions using analysis as presented in [24] (Fig. 4).

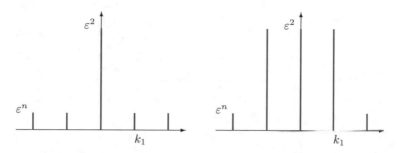

Fig. 4 The mode distribution for $t = 0$ in the KdV case and the mode distribution for $t = \mathcal{O}(|\ln \varepsilon|/\varepsilon^2) \ll \mathcal{O}(1/\varepsilon^3)$. In the NLS case the magnitude ε^2 has to be replaced by ε and the second time is $t = \mathcal{O}(|\ln \varepsilon|/\varepsilon) \ll \mathcal{O}(1/\varepsilon^2)$. The KdV/NLS approximation is no longer valid in the right picture, since the modes at $\pm k_1$ are of the same order as the KdV/NLS modes at $k = 0$

3.2 Resonances in the NLS Case

The error $\varepsilon^\beta R = U - \varepsilon \psi^{\text{NLS}}$, made by the NLS approximation $\varepsilon \psi^{\text{NLS}}$, satisfies

$$\partial_t R = \Lambda R + 2\varepsilon N(\psi^{\text{NLS}}, R) + \mathcal{O}(\varepsilon^2),$$

where $\mathcal{O}(\varepsilon^2)$ contains the nonlinear terms with respect to R and the residual terms, i.e., the terms which do not cancel after inserting the NLS approximation into the BKG system. In order to obtain $\mathcal{O}(\varepsilon^{2+\beta})$ for the residual terms in this equation, higher order terms have to be added to the NLS approximation $\varepsilon \psi^{\text{NLS}}$. This is standard and so we will concentrate on other aspects. Due to the term $2\varepsilon N(\psi^{\text{NLS}}, R)$ a simple application of Gronwall's inequality is not sufficient to obtain an $\mathcal{O}(1)$-bound for R on the long $\mathcal{O}(1/\varepsilon^2)$-time scale. A term $\psi^{\text{NLS}} R_j$ can be eliminated in the i-th equation by a near identity change of variables if the non-resonance condition

$$\omega_i(k) \neq \omega_2(0) + \omega_j(k), \tag{3.5}$$

with $\omega_2(0) = 1$ for (2.1)–(2.2), is satisfied for all $k \in \mathbb{R}$.

The resonances found in Fig. 5 can be used to prove that in case of $2\pi/k_1$-periodicity with $k_2 \notin k_1 \mathbb{N}$ the NLS equation makes wrong predictions about the dynamics of the BKG system. In order to illustrate this we make the ansatz

$$u = \varepsilon^2 A(\varepsilon t) + \varepsilon^n A_1(\varepsilon t) e^{-i\omega_u(k_1)t} e^{ik_1 x} + \text{c.c.},$$

$$v = \varepsilon B(\varepsilon t) e^{it} + \varepsilon^n B_1(\varepsilon t) e^{-i\omega_v(k_1)t} e^{ik_1 x} + \text{c.c.},$$

to analyze the resonance at the wave number $k = k_1$. Equating the coefficients to zero at $\varepsilon e^{it} e^{0ix}$ in the v-equation, at $\varepsilon^n e^{-i\omega_u(k_1)t} e^{ik_1 x}$ in the u-equation, and at

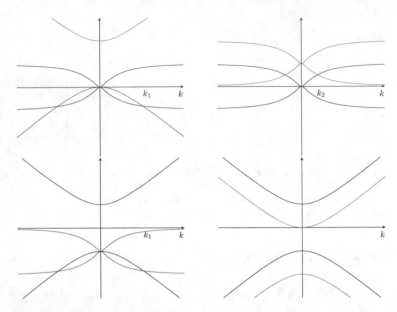

Fig. 5 The intersection points of $k \mapsto \omega_i(k)$, and $k \mapsto \omega_v(0) \pm \omega_j(k)$ correspond to resonances. The associated nonlinear terms cannot be eliminated by near identity change of coordinates. The two graphs in the first line show that in the R_u-equation terms $\psi^{NLS} R_v(k_1)$ and $\psi^{NLS} R_u(k_2)$ for wave numbers k_1 and k_2 cannot be eliminated. For the same wave number k_1 the term $\psi^{NLS} R_u(k_1)$ cannot be eliminated in the R_v-equation

$\varepsilon^n e^{-i\omega_v(k_1)t} e^{ik_1 x}$ in the v-equation, yields, with $\tau = \varepsilon^2 t$, that

$$\partial_\tau^2 A = 0, \tag{3.6}$$

$$2i \partial_\tau B = \mathcal{O}(\varepsilon), \tag{3.7}$$

$$-2i\omega_u(k_1)\partial_\tau A_1 = -2a_{vv}k_1^2 BB_1, \tag{3.8}$$

$$-2i\omega_v(k_1)\partial_\tau B_1 = 2b_{uv}\overline{B}A_1, \tag{3.9}$$

where we used $-\omega_u(k_1) = 1 - \omega_v(k_1)$. The first equation is the NLS equation, i.e., (2.7) restricted to the wave number $k = 0$. Hence, for instance on an $\mathcal{O}(\varepsilon^{-1/2})$-time scale with respect to τ, the variable B can be considered to be constant in time. The last two equations can be written as

$$\partial_\tau^2 A_1 = \Gamma A_1 \qquad \text{resp.} \qquad \partial_\tau^2 B_1 = \Gamma B_1,$$

with

$$\Gamma = \frac{|B|^2}{\omega_u(k_1)\omega_v(k_1)} a_{vv} b_{uv}.$$

Since $\omega_u(k_1)\omega_v(k_1) > 0$, by choosing $a_{vv}b_{uv}$ positive we have growth rates $e^{\beta\tau} = e^{\beta\varepsilon t} = e^{\beta T/\varepsilon}$ with a $\beta > 0$. These allow us to bring $\varepsilon^n A_1$ and $\varepsilon^n B_1$, which are initially of order $\mathcal{O}(\varepsilon^n)$, to an order $\mathcal{O}(\varepsilon)$ at a time $T = \mathcal{O}((n-1)\varepsilon|\ln(\varepsilon)|) \ll 1$. Therefore, we have that $v = \mathcal{O}(\varepsilon)$ far before the natural scale of the NLS equation. Hence, in this situation the NLS approximation makes wrong predictions. These calculations can be transferred into a rigorous proof using analysis as presented in [24].

4 Validity in the Non-oscillatory Case

In this section we discuss the validity of the KdV approximation for the BKG system. There are essentially three different results which we would like to present. As in [1] the subsequent analysis is not only valid for the KdV limit, but also for the inviscid Burgers and the Whitham limit.

4.1 Approach 1: Using Normal Form Transformations in the Non-resonant Case

In [6] the BKG system has been considered in case $\alpha = 1$ or more general in case without additional resonances, i.e., in case $\omega_u(k) \neq \omega_v(k)$ for all $k \in \mathbb{R}$. Then with normal form transformations and energy estimates the following result has been established.

Theorem 4.1 *Let $A \in C([0, T_0], H^8(\mathbb{R}, \mathbb{R}))$ be a solution of the KdV equation (2.5). Then there exist $\varepsilon_0, C > 0$ such that for all $\varepsilon \in (0, \varepsilon_0)$ we have solutions (u, v) of (2.1)–(2.2) with*

$$\sup_{t\in[0,T_0/\varepsilon^3]} \sup_{x\in\mathbb{R}} |(u, v)(x, t) - (\varepsilon^2\psi_u^{\mathrm{KdV}}(x, t), 0)| \leq C\varepsilon^{7/2}.$$

Sketch of the Proof We write a true solution of (2.1)–(2.2) as approximation plus error, i.e., $u = \varepsilon^2\psi_u + \varepsilon^\beta R_u$ and $v = \varepsilon^4\psi_v + \varepsilon^\beta R_v$ with $\beta = 7/2$, where $(\varepsilon^2\psi_u, \varepsilon^4\psi_v)$ is an improved approximation which is formally $\mathcal{O}(\varepsilon^4)$ close to $(\varepsilon^2\psi_u^{\mathrm{KdV}}(x, t), 0)$. The error satisfies

$$\partial_t^2 R_u = \alpha^2\partial_x^2 R_u + \partial_t^2\partial_x^2 R_u + 2\varepsilon^2\partial_x^2(a_{uu}\psi_u R_u + a_{uv}\psi_u R_v) + \mathcal{O}(\varepsilon^3), \quad (4.1)$$

$$\partial_t^2 R_v = \partial_x^2 R_v - R_v + 2\varepsilon^2 b_{uu}\psi_u R_u + 2\varepsilon^2 b_{uv}\psi_u R_v + \mathcal{O}(\varepsilon^3). \quad (4.2)$$

After elimination of the non-resonant terms the system decouples up to order $\mathcal{O}(\varepsilon^3)$, namely

$$\partial_t^2 R_u = \alpha^2 \partial_x^2 R_u + \partial_t^2 \partial_x^2 R_u + 2\varepsilon^2 \partial_x^2 (a_{uu} \psi_u R_u) + \mathcal{O}(\varepsilon^3),$$

$$\partial_t^2 R_v = \partial_x^2 R_v - R_v + 2\varepsilon^2 b_{uv} \psi_u R_v + \mathcal{O}(\varepsilon^3),$$

where we used the same symbols for the old and new variables. Then multiplying the first equation with $\partial_t \partial_x^{-2} R_u$ and the second equation with $\partial_t R_v$ gives, after integration with respect to x, the energy estimates

$$\partial_t \int ((\partial_t \partial_x^{-1} R_u)^2 + \alpha^2 (R_u)^2 + (\partial_t R_u)^2 + 2\varepsilon^2 a_{uu} \psi_u (R_u)^2$$

$$+ (\partial_t R_v)^2 + (\partial_x R_v)^2 + (R_v)^2 - 2\varepsilon^2 b_{uv} \psi_u (R_v)^2) dx = \mathcal{O}(\varepsilon^3),$$

where we used integration by parts, $\partial_t \psi_u = \mathcal{O}(\varepsilon)$, and $\partial_x \psi_u = \mathcal{O}(\varepsilon)$. Hence, the integral on the left hand side stays $\mathcal{O}(1)$-bounded on an $\mathcal{O}(1/\varepsilon^3)$-time scale. Since similar estimates can be obtained for the derivatives, the H^s-norm of the error stays $\mathcal{O}(1)$-bounded on the $\mathcal{O}(1/\varepsilon^3)$-time scale.

4.2   Approach 2: Using the Hamiltonian

In the resonant case the terms $2\varepsilon^2 \partial_x^2 (a_{uv} \psi_u R_v)$ and $2\varepsilon^2 b_{uu} \psi_u R_u$ cannot be eliminated from the equations for the error (4.1)–(4.2). In Sect. 3.1 we have seen that then in case of suitably chosen coefficients positive growth rates occur. There we have also seen that positive growth rates can not occur in case $a_{uv} b_{uu} \leq 0$, i.e., when a_{uv} and b_{uu} have different signs. The second result is obtained in this situation, more precisely, when the lowest order part of the error equation can be written as Hamiltonian system. Then the ideas of [1] apply. There, a first justification result for the KdV approximation of a scalar dispersive PDE, posed in a spatially periodic medium of non-small contrast, has been obtained via some suitably chosen energy. It is based on

$$\frac{d}{dt} H(R(t), t) = \nabla H \cdot \partial_t R(t) + \partial_t H = 0 + \mathcal{O}(\varepsilon^3), \qquad (4.3)$$

since $\varepsilon^2 \partial_t \psi_u = \mathcal{O}(\varepsilon^3)$ due to the long wave character of the KdV approximation with respect to time. The approximation result is as above. The sketch of the proof in this case is as follows. Without performing a normal form transform as before, we multiply the first equation of the system for the error (4.1)–(4.2) with $\partial_t \partial_x^{-2} R_u$

and the second equation with $\partial_t R_v$. This gives after integration with respect to x the energy estimates

$$\partial_t (b_{uu} E_u - a_{uv} E_v) = \varepsilon^2 s_1 + \mathcal{O}(\varepsilon^3),$$

with

$$E_u = \int (\partial_t \partial_x^{-1} R_u)^2 + \alpha^2 (R_u)^2 + (\partial_t R_u)^2 + 2\varepsilon^2 a_{uu} \psi_u (R_u)^2 dx,$$

$$E_v = \int (\partial_t R_v)^2 + (\partial_x R_v)^2 + (R_v)^2 - 2\varepsilon^2 b_{uv} \psi_u (R_v)^2 dx,$$

$$s_1 = 2a_{uv} b_{uu} \int (\partial_t R_u) \psi_u R_v + (\partial_t R_v) \psi_u R_u dx,$$

where we used integration by parts, $\partial_t \psi_u = \mathcal{O}(\varepsilon)$, and $\partial_x \psi_u = \mathcal{O}(\varepsilon)$. Hence, in case of opposite signs of a_{uv} and b_{uu} the term $|b_{uu} E_u - a_{uv} E_v|$ is an energy and for this energy s_1 can be written as time-derivative plus some small error, i.e.,

$$\partial_t \int \psi_u R_u R_v dx + \mathcal{O}(\varepsilon),$$

again due to $\partial_t \psi_u = \mathcal{O}(\varepsilon)$. Therefore, the time derivative term can be included into the energy on the left hand side. Then we have

$$\partial_t (b_{uu} E_u - a_{uv} E_v - 2\varepsilon^2 a_{uv} b_{uu} \int \psi_u R_u R_v dx) = \mathcal{O}(\varepsilon^3),$$

and so the modified energy stays $\mathcal{O}(1)$-bounded on an $\mathcal{O}(1/\varepsilon^3)$-time scale. Since similar estimates can be obtained for the derivatives, the H^s-norm of the error stays $\mathcal{O}(1)$-bounded on the $\mathcal{O}(1/\varepsilon^3)$-time scale.

4.3 Approach 3: Handling Unstable Resonances

The third approach also works in case of unstable resonances. In order to explain the underlying idea we go back to the amplitude system (3.3)–(3.4) describing the unstable resonances. In order to have an $\mathcal{O}(1)$-bound for A_1 on an $\mathcal{O}(1/\varepsilon^3)$-time scale with respect to t we need that A_1 is exponentially small initially, i.e., $A_1(0) = e^{-r/\varepsilon}$ for an $r > 0$. Since $e^{\beta \varepsilon^2 t} e^{-r/\varepsilon} \leq 1$ for $t \leq r/(\beta \varepsilon^3)$ the exponential smallness for $t = 0$ allows us to come at least to the correct time-scale. This idea has to be combined with energy estimates for the wave numbers close to $k = 0$. With this respect the approach is more involved than the one used in [15, 18] for the water wave problem over a flat bottom. There, functions exponentially decaying with respect to the Fourier wave numbers for $|k| \to \infty$ were used for a local existence and uniqueness proof.

Our third approximation result is as follows.

Theorem 4.2 *Let A be a solution of the KdV equation* (2.5) *with*

$$\sup_{T \in [0,T_0]} \int |\widehat{A}(K,T)| e^{r|K|} dK < \infty$$

for an $r > 0$. *Then there exist* $\varepsilon_0 > 0$, $T_1 \in (0, T_0]$, $C > 0$ *such that for all* $\varepsilon \in (0, \varepsilon_0)$ *we have solutions* (u, v) *of* (2.1)–(2.2) *with*

$$\sup_{t \in [0, T_1/\varepsilon^3]} \sup_{x \in \mathbb{R}} |(u, v)(x, t) - (\varepsilon^2 \psi_u^{KdV}(x, t), 0)| \leq C\varepsilon^{7/2}.$$

A detailed proof will be given in a forthcoming paper.

Remark 4.3 For coefficients satisfying $(a_{uu}k_1^2 + b_{uv})^2 - a_{uv}b_{uu}k_1^2 \geq 0$ and $a_{uv}b_{uu} > 0$, the approach of Sect. 4.2 does not apply although we have a stable resonance, cf. Sect. 3.1. We expect that in this parameter regime the approach of Sect. 4.3 with the exponential weights can be avoided and a mixture of normal form transformations and energy estimates like in [20] applies.

5 Validity in the Oscillatory Case

An NLS approximation result in a periodic medium has been obtained in [2]. However, due to $\omega_1(0) = 0$ in the spectral picture plotted in Fig. 2 the approach from [2] does not transfer to the situation we are interested in. A spectral picture, similar to the one for the BKG system, occurs for the Klein–Gordon–Zakharov (KGZ) system. A long wave NLS approximation result for the KGZ system can be found in [17]. A NLS approximation result for wave packets with carrier wave number $k_0 > 0$ for systems including the BKG system can be found in [5, 10, 12]. However, none of these results apply in the situation of long wave NLS approximations with unstable resonances.

In order to explain the underlying idea we again go back the amplitude system (3.8)–(3.9) describing the unstable resonances. In order to have an $\mathcal{O}(1)$-bound for A_1 on an $\mathcal{O}(1/\varepsilon^2)$-time scale with respect to t, we need that A_1 is exponentially small initially, i.e., $A_1(0) = e^{-r/\varepsilon}$ for an $r > 0$. Since $e^{\beta \varepsilon t} e^{-r/\varepsilon} \leq 1$ for $t \leq r/(\beta \varepsilon^2)$ the exponential smallness for $t = 0$ allows us to come at least to the correct time-scale. This idea has to be combined with energy estimates for the wave numbers close to $k = 0$.

Theorem 5.1 *Let A be a solution of the NLS equation* (2.7) *with*

$$\sup_{T \in [0, T_0]} \int |\widehat{A}(K, T)| e^{r|K|} dK < \infty$$

for an r > 0. Then there exist $\varepsilon_0 > 0$, $T_1 \in (0, T_0]$, $C > 0$ such that for all $\varepsilon \in (0, \varepsilon_0)$ we have solutions (u, v) of (2.1)–(2.2) with

$$\sup_{t \in [0, T_1/\varepsilon^2]} \sup_{x \in \mathbb{R}} |(u, v)(x, t) - (0, \varepsilon \psi_u^{NLS}(x, t))| \leq C\varepsilon^{3/2}.$$

A detailed proof will be given in a forthcoming paper.

6 How to Transfer the Ideas to the Water Wave Problem?

In [24] a counterexample has been constructed showing that the NLS approximation makes wrong predictions about the dynamics of the water wave problem with surface tension and periodic boundary conditions, if the surface tension and the periodicity is suitably chosen. Since the water wave problem with a flat bottom is a special case of the periodic bottom case this counterexample transfers to the periodic water wave problem. Since the construction of this counterexample is robust under small perturbations of the bottom b, a counterexample can be constructed for a slightly periodic bottom, too. Therefore, it is the goal of future research to prove theorems similar to Theorems 4.2 and 5.1 for the water wave problem with a periodic bottom. This will be done by controlling the spatially periodic case first, then by handling the case $|l| > 0$ by some perturbation analysis with the help of exponential weights in Bloch space, and finally to use these exponential weights to control the resonances. The linear water wave problem over periodic bottoms has been analyzed in [8]. Spectral gaps in the Bloch wave spectrum have been found.

Acknowledgements The work of Guido Schneider is partially supported by the Deutsche Forschungsgemeinschaft DFG through the SFB 1173 "Wave phenomena".

References

1. R. Bauer, W.-P. Düll, G. Schneider, The Korteweg–de Vries, Burgers and Whitham limits for a spatially periodic Boussinesq model. Proc. R. Soc. Edinburgh Sect. A **149**(1), 191–217 (2019)
2. K. Busch, G. Schneider, L. Tkeshelashvili, H. Uecker, Justification of the nonlinear Schrödinger equation in spatially periodic media. Z. Angew. Math. Phys. **57**(6), 905–939 (2006)
3. F. Chazel, Influence of bottom topography on long water waves. ESAIM Math. Model. Numer. Anal. **41**(4), 771–799 (2007)

4. F. Chazel, On the Korteweg-de Vries approximation for uneven bottoms. Eur. J. Mech. B Fluids **28**(2), 234–252 (2009)
5. M. Chirilus-Bruckner, C. Chong, O. Prill, G. Schneider, Rigorous description of macroscopic wave packets in infinite periodic chains of coupled oscillators by modulation equations. Discrete Contin. Dynam. Syst. Ser. S **5**(5), 879–901 (2012)
6. C. Chong, G. Schneider, The validity of the KdV approximation in case of resonances arising from periodic media. J. Math. Anal. Appl. **383**(2), 330–336 (2011)
7. W. Craig, An existence theory for water waves and the Boussinesq and Korteweg-deVries scaling limits. Commun. Partial Differ. Equ. **10**, 787–1003 (1985)
8. W. Craig, M. Gazeau, C. Lacave, C. Sulem, Bloch theory and spectral gaps for linearized water waves. SIAM J. Math. Anal. **50**(5), 5477–5501 (2018)
9. W.-P. Düll, Validity of the Korteweg-de Vries approximation for the two-dimensional water wave problem in the arc length formulation. Commun. Pure Appl. Math. **65**(3), 381–429 (2012)
10. W.-P. Düll, G. Schneider, Justification of the nonlinear Schrödinger equation for a resonant Boussinesq model. Ind. Univ. Math. J. **55**(6), 1813–1834 (2006)
11. W.-P. Düll, G. Schneider, C.E. Wayne, Justification of the nonlinear Schrödinger equation for the evolution of gravity driven 2D surface water waves in a canal of finite depth. Arch. Ration. Mech. Anal. **220**(2), 543–602 (2016)
12. J. Gaison, S. Moskow, J.D. Wright, Q. Zhang, Approximation of polyatomic FPU lattices by KdV equations. Multiscale Model. Simul. **12**(3), 953–995 (2014)
13. T. Iguchi, A long wave approximation for capillary-gravity waves and an effect of the bottom. Commun. Partial Differ. Equ. **32**(1), 37–85 (2007)
14. L.A. Kalyakin, Asymptotic decay of a one-dimensional wave-packet in a nonlinear dispersive medium. Math. USSR Sb. **60**(2), 457–483 (1988)
15. T. Kano, T. Nishida, A mathematical justification for Korteweg-de Vries equation and Boussinesq equation of water surface waves. Osaka J. Math. **23**, 389–413 (1986)
16. P. Kirrmann, G. Schneider, A. Mielke, The validity of modulation equations for extended systems with cubic nonlinearities. Proc. R. Soc. Edinb. Sect. A **122**(1–2), 85–91 (1992)
17. N. Masmoudi, K. Nakanishi, From the Klein-Gordon-Zakharov system to the nonlinear Schrödinger equation. J. Hyperbolic Differ. Equ. **2**(4), 975–1008 (2005)
18. G. Schneider, Limits for the Korteweg-de Vries-approximation. Z. Angew. Math. Mech. **76**, 341–344 (1996)
19. G. Schneider, Justification of modulation equations for hyperbolic systems via normal forms. NoDEA, Nonlinear Differ. Equ. Appl. **5**(1), 69–82 (1998)
20. G. Schneider, Justification and failure of the nonlinear Schrödinger equation in case of non-trivial quadratic resonances. J. Differ. Equ. **216**(2), 354–386 (2005)
21. G. Schneider, H. Uecker, *Nonlinear PDEs*. Graduate Studies in Mathematics (American Mathematical Society, Providence, 2017). A dynamical systems approach
22. G. Schneider, C.E. Wayne, The long-wave limit for the water wave problem. I: The case of zero surface tension. Commun. Pure Appl. Math. **53**(12), 1475–1535 (2000)
23. G. Schneider, C.E. Wayne, The rigorous approximation of long-wavelength capillary-gravity waves. Arch. Ration. Mech. Anal. **162**(3), 247–285 (2002)
24. G. Schneider, D.A. Sunny, D. Zimmermann, The NLS approximation makes wrong predictions for the water wave problem in case of small surface tension and spatially periodic boundary conditions. J. Dyn. Differ. Equ. **27**(3–4), 1077–1099 (2015)
25. N. Totz, A justification of the modulation approximation to the 3D full water wave problem. Commun. Math. Phys. **335**(1), 369–443 (2015)
26. N. Totz, S. Wu, A rigorous justification of the modulation approximation to the 2D full water wave problem. Commun. Math. Phys. **310**(3), 817–883 (2012)

On Recent Numerical Methods for Steady Periodic Water Waves

Dominic Amann

Abstract The study of steady periodic water waves, analytically as well as numerically, is a very active field of research. We describe some of the more recent numerical approaches to computing these waves numerically as well as the corresponding results. The focus of this work is on the different formulations as well as their limitations and similarities.

Keywords Steady water waves · Numerical methods · Numerical continuation · Nonlocal formulation

1 Introduction

We consider steady water waves in two dimensions, travelling over a flat bottom with speed c and a free surface, under the influence of gravity. This means that in a frame moving along the wave with the same speed c, the velocity field, pressure and shape of the wave does not change over time. This model can be used to study plane waves by considering their cross section perpendicular to the wave crest. For a more detailed derivation of the model equations we refer to [6, 8].

This problem is governed by the Euler equations, find $u(x, y, t)$, $v(x, y, t)$ and $P(x, y, t)$ that solve

$$
\begin{aligned}
u_x + v_y &= 0 & &\text{in } -d < y < \eta(x, t), \\
u_t + uu_x + vu_y &= -P_x & &\text{in } -d < y < \eta(x, t), \\
v_t + uv_x + vv_y &= -P_y - g & &\text{in } -d < y < \eta(x, t)
\end{aligned}
\tag{1}
$$

D. Amann (✉)
Johann Radon Institute for Computational and Applied Mathematics, Linz, Austria
e-mail: dominic.amann@ricam.oeaw.ac.at

© Springer Nature Switzerland AG 2019
D. Henry et al. (eds.), *Nonlinear Water Waves*, Tutorials, Schools, and Workshops in the Mathematical Sciences, https://doi.org/10.1007/978-3-030-33536-6_9

with the free surface $\eta(x, t)$ and the depth d. The fluid domain is sketched on the left hand side of Fig. 1. The boundary conditions for the pressure P and the velocity field (u, v) are the dynamic boundary condition

$$P = P_{\text{atm}} \quad \text{on } y = \eta(x, t) \tag{2}$$

which model the interaction on the free surface where negligible surface tension is assumed. Additionally we have kinematic boundary conditions, those are that the surface of the wave is always made up of the same particles and that the water can not penetrate the flat bottom. These boundary conditions are modelled by

$$v = \eta_t + u\eta_x \quad \text{on } y = \eta(x, t), \tag{3}$$

$$v = 0 \qquad\qquad \text{on } y = -d. \tag{4}$$

Since we consider steady periodic waves, we introduce a frame moving at the constant speed c which removes the time variable from our system. The new system is

$$U_X + V_Y = 0 \qquad\qquad \text{in } -d < Y < \eta,$$

$$(U - c)U_X + VU_Y = -\tilde{P}_X \qquad \text{in } -d < Y < \eta, \tag{5}$$

$$(U - c)V_X + VV_Y = -\tilde{P}_Y - g \quad \text{in } -d < Y < \eta$$

where $(X, Y) = (x - ct, y)$ and $(U, V, \tilde{P})$ are the functions (u, v, P) transformed to the moving frame. The boundary conditions now read

$$\tilde{P} = \tilde{P}_{\text{atm}} \qquad\quad \text{on } Y = \eta, \tag{6}$$

$$V = (U - c)\eta_X \quad \text{on } Y = \eta, \tag{7}$$

$$V = 0 \qquad\qquad \text{on } Y = -d. \tag{8}$$

Of particular interest are rotational waves, that means that the vorticity $\gamma = v_x - u_y$ is not zero. One reason for the importance of vorticity is its influence on the existence and position of stagnation points, these are points in the wave where the velocity of the fluid is equal to the wave speed c. For the effects that stagnation points have on the flow structure of wave see [13, 14, 35]. For example, waves with a non-smooth peak, such as the Stokes wave of maximal height, see [33], have a stagnation point at that peak.

In this work several schemes on how to solve (1)–(4) numerically are discussed. While there exists a large literature concerning this problem, see [10, 16, 29, 30], here some more recent approaches are presented. Section 2 presents two schemes based on a Dubreil-Jacotin transformation, a numerical continuation approach [3, 20, 21, 32] and an asymptotic expansion approach [2, 9, 18]. Section 3 discusses a non-local formulation [1, 4, 11, 15] and a conformal mapping approach [5, 24, 28] is presented in Sect. 4. The schemes have in common that in order to be able to solve their respective systems, they assume all but one of the parameters, like depth, vorticity or velocity, are fixed. The remaining parameter can be varied to continue along the solution branch.

A non-exhaustive list of further numerical schemes not included in this discussion are: in [31] a shape optimisation approach applied to a stream formulation that allows for a non-flat bottom is presented; [17] modifies the nonlinear shallow water equations to allow for constant vorticity and examines wave breaking; the papers [22, 23] study and compute periodic waves based on an integral formulation.

2 Dubreil-Jacotin Transformation

Following the procedure described in [8] we define the stream function ψ by $\psi_x = -V$ and $\psi_y = U - c$. Then the system (5)–(8) can be reformulated as the equivalent system

$$\Delta\psi = \gamma(\psi) \quad \text{in} - d < y < \eta, \tag{9}$$

$$\psi = 0 \quad \text{on } y = \eta, \tag{10}$$

$$|\nabla\psi|^2 + 2g(y + d) - Q\psi = -p_0 \quad \text{on } y = -d, \tag{11}$$

$$\psi = 0 \quad \text{on } y = \eta. \tag{12}$$

Here Q is the hydraulic head, p_0 the relative mass flux and γ the vorticity function. In order to ensure that the vorticity is a function of ψ one has to assume

$$u < c \tag{13}$$

in the whole fluid domain. This condition excludes stagnation points since there it holds $u = c$.

One major difficulty with the original system as well as the stream formulation is the unknown free surface η. In [8, 32] a fixed domain formulation equivalent to (9)–(12) is discussed. The used coordinate transform known as the Dubreil-Jacotin transform, see [12], is illustrated in Fig. 1. This transform exploits that ψ is constant both on the flat bottom and the free surface as well as strictly increasing inside the domain, note that this again makes use of the assumption (13). Introducing the fixed domain $R = \{(q, p) | -\pi < q < \pi; p_0 < p < 0\}$ and the height function $h = y + d$ where y depends implicitly on q and p, results in a new system of equations.

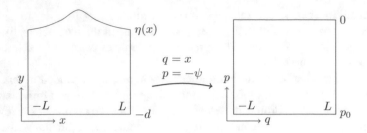

Fig. 1 Dubreil-Jacotin transformation

Hence instead of (9)–(12) the problem is now to find h and Q satisfying

$$\mathcal{H}[h] :=(1 + h_q^2)h_{pp} - 2h_p h_q h_{qp} + h_p^2 h_{qq} - \gamma(-p)h_p^3 = 0 \quad \text{in } R, \qquad (14)$$

$$\mathcal{B}_0[h, Q] := \qquad\qquad\qquad 1 + h_q^2 + (2gh - Q)h_p^2 = 0 \quad \text{for } p = 0, \qquad (15)$$

$$\mathcal{B}_1[h] := \qquad\qquad\qquad\qquad\qquad\qquad h = 0 \quad \text{for } p = p_0 \qquad (16)$$

for a given domain R and vorticity function γ. Due to assumption (13), the formulations presented in this chapter and all schemes based on the Dubreil-Jacotin transformation can not be used to compute waves with stagnation point.

A special family of solutions of these equations are the so called laminar waves. These solutions, defined in the fixed domain R, describe parallel shear flows that do not depend on the q variable and are denoted as H. In the case of linear vorticity the laminar waves are given by

$$H(p; \lambda) = \frac{2(p - p_0)}{\sqrt{\lambda - 2\gamma p} + \sqrt{\lambda - 2\gamma p_0}} \qquad (17)$$

where the parameter $\lambda > 0$ is coupled to Q by the relation D

$$Q = \lambda - \frac{4gp_0}{\sqrt{\lambda} + \sqrt{\lambda - 2\gamma p_0}}.$$

In general, there are no non-laminar waves in the neighbourhood of the laminar wave $(Q(\lambda), H(\lambda))$. For certain values of λ and thus Q, determined by the dispersion relation [19], a branch of non-laminar waves bifurcates from the family of laminar flows. These bifurcation points are denoted as λ_* and Q^* respectively.

While the schemes presented in Sects. 3 and 4 consider different system of equations and computational domains, the concepts of laminar branches and bifurcation of a branch of non-trivial waves remain the same.

2.1 Numerical Continuation Scheme

One straightforward approach is to discretise (14)–(16) with a second order finite difference scheme as was done in [3, 20, 21, 32]. The resulting system is underdetermined since the hydraulic head Q as well as the height function h are unknown. A numerical continuation scheme can be used to compute waves along the solution branch. This means introducing additional conditions to make the system determined and provide initial guesses based on previous solutions, the resulting system of nonlinear equations is solved with a Newton's method.

Finding the bifurcation point Q^* and the initial guess for the first non-trivial wave can be done by either computing eigenvalues of the linearised system, using analytical results [19] or employing other approaches such as the asymptotic expansion, see Sect. 2.2.

For numerical computations, wavelength and the relative mass flux p_0 have to be chosen.[1] In the standard case of a given vorticity function γ the hydraulic head Q is the only free parameter which makes it the natural choice for the bifurcation parameter.

Note that other choices are valid and may be beneficial. For example consider the case of constant vorticity $\gamma = \gamma_0$, then one can fix Q and consider γ_0 as the bifurcation parameter. Given one solution, new waves can then be computed by varying γ_0 while Q remains fixed. This strategy was used in [3] to compute parts of the solution branch beyond a wave with stagnation points. This branch is not reachable by continuation with Q since there, the part of the branch that violates (13), can not be bypassed.

The biggest drawback of this approach is that, due to the assumption (13) of the Dubreil-Jacotin transform, waves with stagnation points are not modelled by (14)–(16). This manifests in an increasingly ill-condition Jacobian of the discretisation near stagnating waves. The advantage of this scheme are its ease of use and the big flexibility it has: no assumptions on the vorticity function γ are made, examples presented in [3, 20, 21] include piecewise constant and cubic vorticity with respect to the stream function; Fig. 2 shows some examples of linear vorticity; The scheme is also flexible with regard to the model equation (1). Extensions, like periodic travelling equatorial waves [7, 27], that add earths rotation to the model, are straightforward to incorporate.

---

[1]A condition for such a choice which ensures existence of solution is given by (1.6) in Theorem 1.1 of [8].

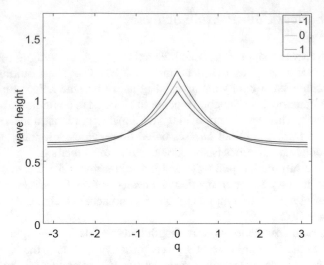

Fig. 2 Free boundary of the limiting wave with vorticity $\gamma_\alpha(p) = \alpha(1 - \frac{p}{p_0})$ for $\alpha \in \{-1, 0, 1\}$

2.2 Asymptotic Expansion

In the approach presented in [2] and based on [9, 18], one considers the fixed domain formulation (14)–(16) and finds asymptotic expansion of its solution around a bifurcation point.

Looking for q-independent solutions of (14)–(16) leads to the laminar flows H. Similarly, one can obtain solutions for the linearised problem with the approach

$$\hat{h}(q, p; b) = H(p, \lambda) + bm(q, p)$$

where $b \in \mathbb{R}$ and m is an even and 2π-periodic function in q. The unknown function m is chosen such that $\hat{h}$ is the solution of the linearised problem, that is

$$\mathcal{H}[\hat{h}] = \mathcal{O}(b^2), \quad \mathcal{B}_0[\hat{h}, Q] = \mathcal{O}(b^2), \quad \mathcal{B}_1[\hat{h}] = 0.$$

This linearised problem only has non-trivial solutions at bifurcation points, that is $(Q^*, H(\lambda_*))$. Those are either known analytically [19] for some vorticities or can be approximated numerically. A way to compute a better approximation of h would be to consider higher order approximations. As discussed in [9], adding more terms to $\hat{h}$ only yields solutions of the system up to second order. There exists no expansion $\hat{h}_3$ of third order that satisfies the system (14)–(16) up to third order, that is

$$\mathcal{H}[\hat{h}_3] = \mathcal{O}(b^4), \quad \mathcal{B}_0[\hat{h}_3, Q] = \mathcal{O}(b^4), \quad \mathcal{B}_1[\hat{h}_3] = 0.$$

This caveat was remedied in [18] by approximating not just the height function h but Q as well. For this introduce approximations of the pair (Q, h) by the polynomials in $b \in \mathbb{R}$

$$Q \approx Q^{(2N)}(b) \qquad := Q^* + \sum_{k=1}^{N} Q_{2k} b^k, \qquad (18)$$

$$h(q, p; Q) \approx h^{(2N+1)}(q, p; b) := \sum_{k=0}^{2N+1} h_k(q, p) b^k \qquad (19)$$

with coefficients $Q_{2k} \in \mathbb{R}$ and h_k defined as

$$h_{2k}(q, p) := \sum_{m=0}^{k} \cos(2mq) f_{2m}^{2k}(p), \qquad (20)$$

$$h_{2k+1}(q, p) := \sum_{m=0}^{k} \cos((2m + 1)q) f_{2m+1}^{2k+1}(p) \qquad (21)$$

where $f_m^k \in C^\infty([p_0, 0])$ for all m, k. Note that the functions f_m^k only depend on p introduced by the Dubreil-Jacotin transformation. Let the wavelength, vorticity γ and relative mass flux p_0 be given. What remains to be computed are the constants Q_{2k} and functions f_m^k such that $(Q^{(2N)}, h^{(2N+1)})$ satisfies (14)–(16) up to $\mathcal{O}(b^{2N+2})$. The structure of $h^{(2N+1)}$, given by (19)–(21), can be exploited to considerably simplify this problem, as shown in [2]. Ultimately, what has to be solved is a series of one dimensional systems of differential equations which can be done numerically.

Due to the used Dubreil-Jacotin transformation, restriction (13) must be satisfied what in turn means this approach is limited to non-stagnation waves. The advantage of this scheme is its flexibility with regard to the vorticity, in particular non constant vorticity is possible, see [2]. This, together with the availability of analytical results for the first couple expansion terms [18], allows the use of this expansion as a very good initial guess for other approaches.

3 Non-local Formulations

In [1], a new, non-local formulation of the Euler equations was presented which is based on the unified transform or Fokas method. While this approach allows for rotational waves, we present here the periodic irrotational case as was considered in [11]. In the irrotational case, that is $\gamma = 0$, the Euler equations can be formulated in terms of a velocity potential ϕ and become

$$\Delta\phi = 0 \qquad\qquad \text{in } -d < y < \eta \qquad (22)$$

$$\phi_y = 0 \qquad\qquad \text{on } y = -d, \qquad (23)$$

$$\eta_t + \phi_x\eta_x = \phi_y \qquad\qquad \text{on } y = \eta, \qquad (24)$$

$$\phi_t + \frac{1}{2}\left(\phi_x^2 + \phi_y^2\right) + g\eta = \sigma\frac{\eta_{xx}}{(1 + \eta_x^2)^{3/2}} \qquad \text{on } y = \eta \qquad (25)$$

where σ denotes the constant surface tension and ρ is the density.

Introduce the velocity potential evaluated at the surface, see [36], as $q(x,t) = \phi(x, \eta(x,t), t)$ which leads to the dynamic boundary condition

$$q_t + \frac{1}{2}q_x^2 + g\eta - \frac{1}{2}\frac{(\eta_t + q_x\eta_x)^2}{1 + \eta_x^2} = \sigma\frac{\eta_{xx}}{(1 + \eta_x^2)^{3/2}}. \qquad (26)$$

Additionally one gets a non-local equation

$$\int_0^L e^{-ikx}\{i\eta_t\cosh[k(\eta + d)] + q_x\sinh[k(\eta + d)]\}dx = 0, \quad t > 0 \qquad (27)$$

where $k = k_n = \frac{2k\pi}{L}$ with $n \in \mathbb{Z} \setminus \{0\}$.

Starting from the set of Eqs. (26)–(27) several modifications and generalisations can be made. Considerations include the constant vorticity case [4, 34], a variable bottom [1] and using a moving frame [11]. A hybrid of the novel formulation and an approach based on conformal mapping is presented in [15], where water waves with variable bottom are studied numerically.

For numerical considerations in the case of steady periodic water waves (26) and (27) can be reformulated as a single non-local equation only containing the unknown η, see [11]. The wave profile η is approximated by truncated Fourier series and the non-local equation discretised using a spectral collocation method. Then the problem to find solutions can be seen as a bifurcation problem for fixed wavelength and depth where the wave velocity c is the bifurcation parameter. To find a bifurcation point for which non-trivial solutions exist, the null space of the linearisation about the trivial wave is studied.

This approach allows for the computation of streamlines and pressure in the whole domain, independent of any grid. This holds true even in the presence of stagnation points when rotational waves are considered. For example, in [34] a wave with interior stagnation and a bottom pressure maxima which is not under the crest is presented.

Such non-local formulations are a very active research area, for some more related formulations see [25, 26, 34]. This, together with the easily available information about streamlines, wave form and pressure, make non-local formulations very effective. The main limitation is that the vorticity function has a larger impact on the formulation and is thus more restricted, in most cases to constant vorticity.

4 Conformal Mapping and Spectral Collocation Method

In the approach presented in [28], the constant vorticity case is considered as a superposition of a linear shear flow and a harmonic velocity potential. This leads to a system of equations similar to (22)–(25) but with an additional vorticity term, which is then non-dimensionalised. The manuscript [28] considers the case of periodic travelling waves with constant speed and introduces a frame moving along with wave speed c. Then the problem is to find the potential ϕ satisfying

$$\Delta\phi = 0 \quad \text{in } -1 < y < \eta \quad (28)$$

$$\phi_y = 0 \quad \text{on } y = -1, \quad (29)$$

$$-c\eta_x + (\phi_x - \gamma(\eta + b))\eta_x = \phi_y \quad \text{on } y = \eta, \quad (30)$$

$$-c\phi_x + \frac{1}{2}\left(\phi_x^2 + \phi_y^2\right) + \eta - \gamma(\eta + b)\phi_x + \gamma\psi = B \quad \text{on } y = \eta \quad (31)$$

where B is the Bernoulli constant, ψ is the streamfunction associated with ϕ and $b \in \mathbb{R}$ is a parameter of the background flow.

To solve this system, a conformal mapping such as given in [5, 24], that maps the uniform strip onto the wave domain, is considered. In the uniform strip domain, a flat domain of unknown depth, the solution of the Laplace equation is analytically known. It is ensured that this solution satisfies the dynamic and kinematic boundary conditions using a spectral collocation method. For a sketch of the involved domains see Fig. 1, where reversely the fluid domain was mapped onto a rectangle domain.

To compute a first non-laminar wave the irrotational case of small amplitude, for which good approximations are available, is considered. More waves along the solution branch can be computed using a continuation scheme with previous solution as initial guess. For continuation parameters, [28] studies two cases. In the first case, the depth and wave height H are fixed and the continuation parameter is the wavelength λ. In the second case, the depth and wavelength are fixed and either the vorticity γ or the steepness parameter $\frac{H}{\lambda}$ are varied.

This approach can be used to compute waves with stagnation points as well as wave characteristics such as streamlines, stagnation points, particle paths and the pressure. The results presented in [28] include waves with up to three interior stagnation points and waves with switched pressure maxima and minima at the bottom opposed to the irrotational case. The various continuation schemes allow for the detailed study of interactions between parameters and wave characteristics. The biggest restriction of this approach is that it is limited to constant vorticity.

Acknowledgements The author was supported by the project *Computation of large amplitude water waves* (P 27755-N25), funded by the Austrian Science Fund (FWF). The author would like to thank the reviewers for their suggestions and comments, as those led to a more coherent and precise manuscript.

References

1. M. Ablowitz, A.S. Fokas, Z. Musslimani, On a new non-local formulation of water waves. J. Fluid Mech. **562**, 313–343 (2006)
2. D. Amann, K. Kalimeris, Numerical approximation of water waves through a deterministic algorithm. J. Math. Fluid Mech. **20**, 1815–1833 (2018)
3. D. Amann, K. Kalimeris, A numerical continuation approach for computing water waves of large wave height. Eur. J. Mech. B/Fluids **67**, 314–328 (2018)
4. A.C. Ashton, A. Fokas, A non-local formulation of rotational water waves. J. Fluid Mech. **689**, 129–148 (2011)
5. W. Choi, Nonlinear surface waves interacting with a linear shear current. Mat. Comput. Simul. **80**(1), 29–36 (2009)
6. A. Constantin, *Nonlinear Water Waves with Applications to Wave-Current Interactions and Tsunamis*, vol. 81 (SIAM, Philadelphia, 2011)
7. A. Constantin, On the modelling of equatorial waves. Geophys. Res. Lett. **39**(5), L05602 (2012)
8. A. Constantin, W. Strauss, Exact steady periodic water waves with vorticity. Commun. Pure Appl. Math. **57**(4), 481–527 (2004)
9. A. Constantin, K. Kalimeris, O. Scherzer, Approximations of steady periodic water waves in flows with constant vorticity. Nonlinear Anal. Real World Appl. **25**, 276–306 (2015)
10. A.T. Da Silva, D. Peregrine, Steep, steady surface waves on water of finite depth with constant vorticity. J. Fluid Mech. **195**, 281–302 (1988)
11. B. Deconinck, K. Oliveras, The instability of periodic surface gravity waves. J. Fluid Mech. **675**, 141–167 (2011)
12. M.-L. Dubreil-Jacotin, Sur la détermination rigoureuse des ondes permanentes périodiques d'ampleur finie. J. Math. Pures Appl. **13**, 217-291 (1934)
13. M. Ehrnström, J. Escher, E. Wahlén, Steady water waves with multiple critical layers. SIAM J. Math. Anal. **43**(3), 1436–1456 (2011)
14. M. Ehrnström, J. Escher, G. Villari, Steady water waves with multiple critical layers: interior dynamics. J. Math. Fluid Mech. **14**(3), 407–419 (2012)
15. A.S. Fokas, A. Nachbin, Water waves over a variable bottom: a non-local formulation and conformal mappings. J. Fluid Mech. **695**, 288–309 (2012)
16. D. Henry, Large amplitude steady periodic waves for fixed-depth rotational flows. Commun. Partial Differ. Equ. **38**(6), 1015–1037 (2013)
17. V.M. Hur, Shallow water models with constant vorticity. Eur. J. Mech. B/Fluids **73**, 170–179 (2019)
18. K. Kalimeris, Asymptotic expansions for steady periodic water waves in flows with constant vorticity. Nonlinear Anal. Real World Appl. **37**, 182–212 (2017)
19. P. Karageorgis, Dispersion relation for water waves with non-constant vorticity. Eur. J. Mech. B/Fluids **34**, 7–12 (2012)
20. J. Ko, W. Strauss, Effect of vorticity on steady water waves. J. Fluid Mech. **608**, 197–215 (2008)
21. J. Ko, W. Strauss, Large-amplitude steady rotational water waves. Eur. J. Mech. B/Fluids **27**(2), 96 – 109 (2008)
22. M. Longuet-Higgins, M. Tanaka, On the crest instabilities of steep surface waves. J. Fluid Mech. **336**, 51–68 (1997)
23. D.V. Maklakov, Almost-highest gravity waves on water of finite depth. Eur. J. Appl. Math. **13**(1), 67–93 (2002)
24. P.A. Milewski, J.-M. Vanden-Broeck, Z. Wang, Dynamics of steep two-dimensional gravity–capillary solitary waves. J. Fluid Mech. **664**, 466–477 (2010)
25. K. Oliveras, V. Vasan, A new equation describing travelling water waves. J. Fluid Mech. **717**, 514–522 (2013)

26. K.L. Oliveras, V. Vasan, B. Deconinck, D. Henderson, Recovering the water-wave profile from pressure measurements. SIAM J. Appl. Math. **72**(3), 897–918 (2012)
27. R. Quirchmayr, On irrotational flows beneath periodic traveling equatorial waves. J. Math. Fluid Mech. **19**(2), 283–304 (2017)
28. R. Ribeiro, P.A. Milewski, A. Nachbin, Flow structure beneath rotational water waves with stagnation points. J. Fluid Mech. **812**, 792–814 (2017)
29. L. Schwartz, J. Fenton, Strongly nonlinear waves. Annu. Rev. Fluid Mech. **14**(1), 39–60 (1982)
30. J.A. Simmen, P. Saffman, Steady deep-water waves on a linear shear current. Stud. Appl. Math. **73**(1), 35–57 (1985)
31. M. Souli, J. Zolesio, A. Ouahsine, Shape optimization for non-smooth geometry in two dimensions. Comput. Methods Appl. Mech. Eng. **188**(1), 109–119 (2000)
32. W. Strauss, Steady water waves. Bull. Am. Math. Soc. **47**(4), 671–694 (2010)
33. J.F. Toland, On the existence of a wave of greatest height and stokes's conjecture. Proc. R. Soc. Lond. A Math. Phys. Sci. **363**(1715), 469–485 (1978)
34. V. Vasan, K. Oliveras, Pressure beneath a traveling wave with constant vorticity. Discrete Contin. Dynam. Syst. **34**, 3219–3239 (2014)
35. E. Wahlén, Steady water waves with a critical layer. J. Differ. Equ. **246**(6), 2468–2483 (2009)
36. V.E. Zakharov, Stability of periodic waves of finite amplitude on the surface of a deep fluid. J. Appl. Mech. Tech. Phys. **9**(2), 190–194 (1968)

Nonlinear Wave Interaction in Coastal and Open Seas: Deterministic and Stochastic Theory

Raphael Stuhlmeier, Teodor Vrecica, and Yaron Toledo

Abstract We review the theory of wave interaction in finite and infinite depth. Both of these strands of water-wave research begin with the deterministic governing equations for water waves, from which simplified equations can be derived to model situations of interest, such as the mild slope and modified mild slope equations, the Zakharov equation, or the nonlinear Schrödinger equation. These deterministic equations yield accompanying stochastic equations for averaged quantities of the sea-state, like the spectrum or bispectrum. We discuss several of these in depth, touching on recent results about the stability of open ocean spectra to inhomogeneous disturbances, as well as new stochastic equations for the nearshore.

Keywords Water waves · Nonlinear interaction · Kinetic equations · Shoaling · Zakharov equation · Nonlinear Schrödinger equation · Mild-slope equation · Wave forecasting · Deep water · Nearshore · Resonant interaction

Mathematics Subject Classification (2000) Primary 76B15; Secondary 86A05

1 Introduction

1.1 Preliminaries

The water wave problem as it is understood today is an outgrowth of Newtonian mechanics, and was first cast in the framework of partial differential equations by Leonhard Euler. From its very beginnings, the development of water wave theory

R. Stuhlmeier (✉)
Centre for Mathematical Sciences, University of Plymouth, Plymouth, UK
e-mail: raphael.stuhlmeier@plymouth.ac.uk

T. Vrecica · Y. Toledo
School of Mechanical Engineering, Tel-Aviv University, Tel-Aviv, Israel
e-mail: teodorv@post.tau.ac.il; toledo@tau.ac.il

© Springer Nature Switzerland AG 2019
D. Henry et al. (eds.), *Nonlinear Water Waves*, Tutorials, Schools, and Workshops in the Mathematical Sciences, https://doi.org/10.1007/978-3-030-33536-6_10

went hand in hand with the development of new mathematical tools for treating differential equations. Belying its classical origins, the subject of water waves remains a vibrant area of research to this day.

Much applicable research on ocean waves today is focused on forecasting, which adds a stochastic element to the deterministic equations for the free boundary problem for an inviscid, incompressible fluid. The purpose of the present review is to present some of these ideas, as well as some recent developments, in the subject of deterministic and stochastic wave interaction. Far from being a mathematical abstraction, this body of ideas informs the surfer waiting for a big swell, the marine engineer designing an offshore structure, and the commercial mariners voyaging across the world's oceans and seas.

1.2 *Governing Equations*

The governing equations for water waves can by now be found in any textbook on the subject. A clear, modern derivation may be found in Johnson [37]. In what follows, the assumptions made of the water will be as follows: it is inviscid (to avoid the Navier–Stokes equations), it is incompressible (so the speed of sound is infinite), and the only restoring force is gravity. The surface tension of water plays an important role for very short waves (periods less than about half a second), but on these scales viscosity also becomes important, and it is expedient to dispense with both. Usually only a single fluid (the water) is considered, and the air above is neglected, allowing a decoupling of the atmosphere from the ocean. This assumption is realistic for the propagation of ocean waves without wind, but must be viewed critically when wind forcing becomes important. One final assumption, less convincing on purely physical grounds, but mathematically important, is that of irrotational flow. While Kelvin's circulation theorem may be invoked to justify this choice, the mathematical convenience of potential flow, i.e. replacing the fluid velocity field $\mathbf{u}$ by a potential ϕ with $\nabla \phi = \mathbf{u}$, is critical in simplifying all subsequent analysis.

The governing equations with these assumptions are as follows:

$$\Delta \phi = 0 \tag{1.1}$$

$$\eta_t + \phi_x \eta_x + \phi_y \eta_y - \phi_z = 0 \text{ on } z = \eta(x, y, t) \tag{1.2}$$

$$\phi_t + \frac{1}{2} \left(\phi_x^2 + \phi_y^2 \right) + g\eta = 0 \text{ on } z = \eta(x, y, t) \tag{1.3}$$

$$\phi_z + \phi_x h_x + \phi_y h_y = 0 \text{ on } z = -h(x, y) \tag{1.4}$$

Here g is the (constant) acceleration of gravity, h denotes the bottom boundary, and η the unknown free-surface. While the bottom may be allowed to vary in space, it will be fixed in time—so we cannot consider, for example, the generation of a tsunami by an earthquake.

Since all nonlinearity is contained in the kinematic surface (1.2) and bottom (1.4) boundary conditions, and the Bernoulli condition (1.3), upon linearization this problem becomes a standard exercise in solving Laplace's equation.

Allowing for bathymetry, the linear problem takes the form

$$\Delta\phi = 0, \tag{1.5}$$

$$\phi_{tt} + g\phi_z = 0 \text{ on } z = 0, \tag{1.6}$$

$$\phi_z + \phi_x h_x + \phi_y h_y = 0 \text{ on } z = -h(x, y), \tag{1.7}$$

where the surface boundary conditions have been combined to eliminate η from the problem. Equation (1.7) simplifies to $\phi_z = 0$ on $z = -h$ for constant depth, resulting in the Laplace equation on a horizontal strip, or, for infinite depth, a half space.

It suffices here to record a few main results: travelling wave solutions in constant depth have the form $\exp(i(\mathbf{kx} - \omega(\mathbf{k})t))$, where the relationship between $\mathbf{k}$ and $\omega(\mathbf{k})$ is given by

$$\omega(\mathbf{k})^2 = g|\mathbf{k}| \tanh(|\mathbf{k}|h). \tag{1.8}$$

When the depth varies it makes sense to define a local wavenumber and frequency— in general we may have $\mathbf{k} = \mathbf{k}(\mathbf{x}, t)$, $\omega = \omega(\mathbf{x}, t)$. Thus, we have travelling wave solutions $\exp(iS(\mathbf{x}, t))$ for a phase-function S, and through this define $\mathbf{k} = \nabla S$ and $\omega = \partial S / \partial t$.

2 Nonlinear Waves and Interaction

The theory of nonlinear water waves was historically first treated by perturbation expansions, dating back to the work of Stokes in the mid nineteenth century. The procedure starts by expanding ϕ and η in (1.1)–(1.4) in terms of a small factor ε, and transferring the free surface from $z = \varepsilon\eta$ to $z = 0$ by a Taylor expansion. One may then solve (1.5)–(1.7) first for terms of order $O(1)$, the solution of which then appears as an inhomogeneity in the equations for $O(\varepsilon^1)$, and so on. The algebra quickly becomes cumbersome, particularly for finite depth, and if more than one wave-train is involved.

It is easier to start with simpler equations, and a good introduction is furnished by Whitham [74, Sec. 15.6]. Assume for the moment that we have a nonlinear, dispersive equation of the form

$$\phi_{tt} + \mathcal{L}_x(\phi) = \varepsilon\mathcal{N}(\phi),$$

where ε is some small parameter, $\mathcal{L}_x$ is a linear differential operator involving spatial (x) derivatives, and $\mathcal{N}$ is some nonlinear operator. The linearised problem, for $\varepsilon =$

0, has travelling wave solutions of the form $\exp(i(kx - \omega(k)t))$, as above, where ω depends on the operator $\mathcal{L}_x$.

In the linear problem, the sum of two plane-wave solutions $\exp(i(k_1 x - \omega_1 t))$ and $\exp(i(k_2 x - \omega_2 t))$ is again a solution. However, if $\mathcal{N}$ contains a term ϕ^2, then a sum of two solutions results in the product $\exp(i((k_1 + k_2)x - (\omega_1 + \omega_2)t))$ on the right-hand side. If the nonlinearity is cubic, then three travelling waves can combine on the right-hand side to $\exp(i((k_1 + k_2 + k_3)x - (\omega_1 + \omega_2 + \omega_3)t))$. These terms act as a forcing for the linear equation $\phi_{tt} + \mathcal{L}_x(\phi)$. Just as in the classical theory of forced linear oscillators, the critical phenomenon is *resonance,* when the frequency of the forcing matches that of the unforced system.

Accounting for all possible sums and differences, we see that resonances for quadratic nonlinearities involve three waves (one from the left-hand side of the equation, and two from the right)

$$
\begin{cases}
\pm \mathbf{k}_1 \pm \mathbf{k}_2 \pm \mathbf{k}_3 = 0, \\
\pm \omega(\mathbf{k}_1) \pm \omega(\mathbf{k}_2) \pm \omega(\mathbf{k}_3) = 0,
\end{cases}
\tag{2.1}
$$

and those for cubic nonlinearities involve four waves

$$
\begin{cases}
\pm \mathbf{k}_1 \pm \mathbf{k}_2 \pm \mathbf{k}_3 \pm \mathbf{k}_4 = 0, \\
\pm \omega(\mathbf{k}_1) \pm \omega(\mathbf{k}_2) \pm \omega(\mathbf{k}_3) \pm \omega(\mathbf{k}_4) = 0.
\end{cases}
\tag{2.2}
$$

These very relations also arise in the expansion of the water wave problem in a small parameter (like the wave slope ka), where the dispersion relation is given by (1.8). In the limit of infinite depth ($h \to \infty$) (1.8) reduces to $\omega(\mathbf{k}) = \sqrt{g|\mathbf{k}|}$, and (2.1) cannot be fulfilled nontrivially. For this limit, (2.2) can be fulfilled only for two $+$ and two $-$ signs in both equations.

The opposite extreme, of shallow water ($|\mathbf{k}|h \ll 1$) means that $\omega = |\mathbf{k}|\sqrt{gh}$, whereby already (2.1) can be fulfilled, provided the signs are not all the same and all wave components propagate in the same direction (see Fig. 1c and d). Due to the lack of dispersivity (i.e. all wave frequencies propagate with the same celerity) (2.2) is also fulfilled, as are the resonance conditions for any higher order interaction (see [73]). Nevertheless, the evolution equations are almost always limited to $O(ka)^2$ for finite water depth. While higher order nonlinear terms may be relevant for high wave steepness, their treatment is extremely cumbersome and will not be considered further.

As waves propagate from deep to shallow water, they are transformed due to bottom changes. In intermediate waters, the changing depth induces a change of the wavenumbers through the dispersion relation (1.8). This alone does not enable the closure of a triad resonance condition and an additional component is required. This component may be supplied by a bottom-induced free-surface interference, which does not satisfy the dispersion relation.

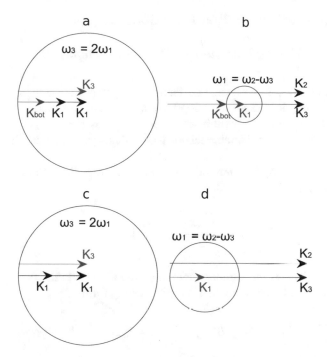

Fig. 1 Resonance conditions in intermediate (**a** and **b**) and shallow water (**c** and **d**). Superharmonic (**a** and **c**), and subharmonic interactions are shown (**b** and **d**). The $\mathbf{k}_{bot}$ represents the bottom component which closes the Bragg III resonance condition

Assume the bottom profile h is decomposed into a sum of sinusoids, then any bottom wavelength can act as a fourth wavenumber component ($\mathbf{k}_{bot}$) with a still ($\omega(\mathbf{k}_{bot}) = 0$) disturbance on the surface. This allows the closure of (2.2) in what is known as a class III Bragg-type resonance condition

$$\begin{cases} \pm\mathbf{k}_1 \pm \mathbf{k}_2 \pm \mathbf{k}_3 = \pm\mathbf{k}_{bot} \\ \pm\omega(\mathbf{k}_1) \pm \omega(\mathbf{k}_2) \pm \omega(\mathbf{k}_3) = 0, \end{cases} \quad (2.3)$$

with a triad of waves. This closure can be represented graphically. Figure 1a–d shows this closure for superharmonic and subharmonic 1D interactions in intermediate water and shallow water conditions, respectively. In the 2D case, bottom components can close resonance with any direction of the third wave $\mathbf{k}_3$. The circles represent the wavenumber $\mathbf{k}_3$ satisfying the closure in all possible directions. Depending on its direction, a bottom component $\mathbf{k}_{bot}$ that satisfies the class III closure should connect its origin on the circle with the origin of the other two waves (see [43, Sec. 3.3]).

3   Deterministic Evolution Equations

3.1   The Zakharov Equation for Constant Depth

It is expedient to take a consistent Fourier analysis perspective with Eqs. (1.1)–(1.4), under the assumption of constant depth h, rather than considering interaction of individual wave-trains via perturbation theory, as was done in [45, 54, 70]. This approach leads after considerable labour to the Zakharov equation, first derived in [77].

In terms of a complex amplitude $B(\mathbf{k}, t)$, related at lowest order to the free-surface elevation η by

$$\hat{\eta}(\mathbf{k}, t) = \sqrt{\frac{\omega(\mathbf{k})}{2g}} \left(B(\mathbf{k}, t)e^{i\omega(\mathbf{k})t} + \text{c.c.} \right) \tag{3.1}$$

(where "c.c." denotes the complex conjugate, and $\hat{\ }$ denotes the $(\mathbf{x} \to \mathbf{k})$ Fourier transform) the Zakharov equation has the following form

$$i\frac{\partial B(\mathbf{k}, t)}{\partial t} = \iiint_{-\infty}^{\infty} T_{0123} B_1^*(t) B_2(t) B_3(t) \delta_{0+1-2-3} e^{i(\Delta_{0+1-2-3})t} \, d\mathbf{k}_1 d\mathbf{k}_2 d\mathbf{k}_3 \tag{3.2}$$

where $\delta_{0+1-2-3} = \delta(\mathbf{k} + \mathbf{k}_1 - \mathbf{k}_2 - \mathbf{k}_3)$ is the delta-distribution, $\Delta_{0+1-2-3} = \omega(\mathbf{k}) + \omega(\mathbf{k}_1) - \omega(\mathbf{k}_2) - \omega(\mathbf{k}_3)$, and $T_{0123} = T(\mathbf{k}, \mathbf{k}_1, \mathbf{k}_2, \mathbf{k}_3)$ is a very lengthy interaction kernel (see [42]). For brevity we have denoted by $B_j(t)$ the complex amplitude $B(\mathbf{k}_j, t)$, and the superscript $*$ denotes a complex conjugate. While a detailed discussion of the Zakharov equation can be found elsewhere (see, e.g. [48, Sec. 14]) it is important to appreciate that Eq. (3.2) arises from a multiple-scale *ansatz* for the full third-order Fourier-space problem, and so captures terms with cubic nonlinearities. In terms of the small wave steepness ε, the time t in this equation is related to physical time T by $t = \varepsilon^2 T$, which is the same as the slow time-scale for the evolution of the envelope in the nonlinear Schrödinger equation [37, Eq. (4.2)]. This long time scale must be borne in mind for all subsequent results.

For computational implementation (and even analytic studies of systems with few waves) it is inevitable that (3.2) must be discretized. Making an ansatz that $B(\mathbf{k}, t) = \sum_{i=1}^{N} B_i(t)\delta(\mathbf{k} - \mathbf{k}_i)$, i.e. that the complex amplitudes can be written as a sum of generalized functions, and integrating over a ball centered around $\mathbf{k}_n$ yields

$$i\frac{dB_n(t)}{dt} = \sum_{p,q,r=1}^{N} T_{npqr}\delta_{n+p-q-r} e^{i\Delta_{n+p-q-r}t} B_p^*(t) B_q(t) B_r(t) \tag{3.3}$$

where

$$\delta_{n+p-q-r} = \begin{cases} 1 \text{ for } \mathbf{k}_n + \mathbf{k}_p = \mathbf{k}_q + \mathbf{k}_r \\ 0 \text{ otherwise} \end{cases} \tag{3.4}$$

now denotes a Kronecker delta-function. Note that other approaches to discretization have been taken, for example by Rasmussen and Stiassnie [57] or Gramstad et al. [26], although their applications largely await further study.

The simplest solution to (3.3) is obtained when only a single wave is present, where with the identity (valid for deep-water)

$$T(\mathbf{k}, \mathbf{k}, \mathbf{k}, \mathbf{k}) = \frac{|\mathbf{k}|^3}{4\pi^2} \tag{3.5}$$

the third-order Stokes' correction is recovered as expected (see [48, Sec. 14.5]). Similarly, if only two collinear waves are present, and we denote these by scalar wavenumbers k_1, k_2 the identity

$$T(k_1, k_2, k_1, k_2) = \begin{cases} k_1 k_2^2/(4\pi^2) \text{ if } k_2 < k_1 \\ k_1^2 k_2/(4\pi^2) \text{ if } k_2 \geq k_1 \end{cases} \tag{3.6}$$

(see [78, Eq. (4.18)]) can be used to establish the mutual Stokes' correction of two wave-trains in deep water, as in [45, eq. (2.11)]. In finite depth the kernels are more involved, and do not yield such compact expressions. In particular, work of Janssen and Onorato [35] first pointed out the problem of non-unique limits for the finite-depth kernel $T(\mathbf{k}, \mathbf{k}, \mathbf{k}, \mathbf{k})$, which was later studied in depth, including for kernels of the form $T(\mathbf{k}, \mathbf{k}_1, \mathbf{k}, \mathbf{k}_1)$, by Stiassnie and Gramstad [63] and Gramstad [24]. A significant consequence of [35] is that modulational instability was shown to disappear for $k_0 h < 1.363$, where k_0 is the carrier wavenumber, and h the (constant) water depth.

In fact, in Zakharov's [77, Eq. (2.3)] derivation of the eponymous equation, it was not the endpoint of his analysis, but rather a step towards the derivation of the nonlinear Schrödinger equation (NLS), which itself was used to study the stability of deep-water waves to perturbations. Having moved from a description in physical variables (x, y, z, t) of fluid motion via the PDEs (1.1)–(1.4), to a third-order simplification (written in terms of variables defined only at the free surface, thus eliminating z) in (k_x, k_y, t) via the integro-differential Zakharov equation (3.2), it is possible to make further restrictions so that an inverse Fourier-transform can be carried out.

The central assumption needed to derive NLS is that all interacting waves are clustered about a single wavenumber, say $\mathbf{k}_0$, an assumption usually referred to as "narrow-bandwidth". This is less apparent when deriving the NLS from the governing equations by perturbation theory [37, Sec. 4.1.1], but implicit also in this formulation. On this basis, the Zakharov kernel in (3.2) is replaced by the kernel

$T(\mathbf{k}_0, \mathbf{k}_0, \mathbf{k}_0, \mathbf{k}_0)$ of (3.5), and the frequency $\omega(\mathbf{k})$ is expanded in a Taylor series about $\omega(\mathbf{k}_0)$. These two steps allow the inverse Fourier transform to be carried out, and lead to the NLS in much the same way that Zakharov [77, Eq. (2.7)ff] first outlined.

3.2   Shallow Water and Varying Depth

We have noted in Sect. 2 that no triad resonance is possible in finite, constant depth. The nature of the perturbation arguments involved implies that quadratic terms are associated with faster time-scales (and larger corrections) than cubic terms, which are in turn more significant than quartic terms—assuming all the necessary interactions are allowed by the dispersion relation. Thus the Zakharov equation (3.2) contains only cubic terms, the non-resonant quadratic terms having been eliminated, and the resonant quartic terms being neglected (indeed, it would be more accurate to refer to (3.2) as the reduced Zakharov equation, see [42] or [64] for the related fourth-order equations in constant depth). For waves in the deep water of the open ocean, this is perfectly satisfactory, but once waves enter coastal environments new equations are needed to capture the effects of a changing bathymetry.

In the shallow water limit, waves become non-dispersive and are able to close exact triad resonances. In real seas waves will almost always tend to steepen and break before reaching the shallow water limit allowing for exact resonance. Nevertheless, breaking does not extract all of the wave energy immediately. It is a gradual process, in which breaking and nonlinearity are coupled. Due to the inherent complexities of wave breaking (only empirical terms for breaking exist), we simply note its importance and include a general dissipation term in the equations. Hence, no nonlinear shoaling examples that include breaking are presented in this chapter. However, even without reaching exact resonance, the nonlinear triad interactions are still of great importance in coastal areas. It is shown here that wave propagation even over mildly varying bathymetry [43] leads to quasi-resonance and significant transformations of wave spectra. The subsequent deterministic model equations are often called mild slope-type equations [4].

The derivation of the linear mild slope equation is based on Eqs. (1.5)–(1.7). If the bed is flat, i.e. h is constant, the Laplace equation can be separated, and the vertical component of the velocity potential is

$$f(z) = \frac{\cosh k(z + h)}{\cosh kh}. \tag{3.7}$$

The key to the mild slope equation is assuming this functional form for the z-dependent part of the solution, even when the bottom is not of constant depth. Thus the solution to (1.5)–(1.7) is written $\phi(x, y, z) = -ig\eta(x, y)\omega^{-1}f(z)$, (here ϕ denotes the time-harmonic velocity potential) and the explicit depth-dependence is integrated out using Green's identity:

$$\int_{-h}^{0} (f_{zz}\phi - f\phi_{zz})\,\mathrm{d}z = [f_z\phi - f\phi_z]_{-h}^{0}.$$

For a flat bed $-h$ = const. the right-hand side vanishes, and we are left with a Helmholtz equation. Otherwise, a varying bed $h = h(x, y)$ gives rise to the mild slope equation when terms $O(\nabla_h h)^2$ and $O(\nabla_h^2 h)$ are neglected:

$$\nabla_h \cdot (a\nabla_h\hat\phi) + k^2 a\hat\phi = 0, \qquad (3.8)$$

where $a(x, y) = g \int_{-h(x,y)}^{0} f(z)^2 dz$. Here, $\nabla_h = (\partial_x, \partial_y)$ is the horizontal gradient and $\hat\phi$ is a single harmonic of the velocity potential on the linearised free surface $(z = 0)$. In fact, a is exactly the product of the phase velocity and the group velocity, $a = \omega/k \cdot d\omega/dk = C_p \cdot C_g$. More details can be found in [48, Ch. 3.5]. Note that upon retaining higher order bottom terms one can derive the more accurate Modified MSE (see [14]).

Nonlinear mild-slope evolution equations can also be derived from the governing equations (1.1)–(1.4), with the vertical structure of velocity potential either assumed to be that of a free wave as in Eq. (3.7) [4, 38], or expanded as a Frobenius series [13, 73], with the latter giving better accuracy in the nonlinear part. The general form, written in terms of the surface velocity potential for a given harmonic p is defined as

$$\nabla_h^2\hat\phi_p + \frac{\nabla_h (C_p C_{g,p}) \cdot \nabla_h\hat\phi_p}{C_p C_{g,p}} + k_p^2\hat\phi_p = \mathrm{NL}_p, \qquad (3.9)$$

where C_p and $C_{g,p}$ are wave celerity and group velocity for harmonic p, while NL_p is the nonlinear triad term, which closes an exact resonance in frequency for harmonic p.

In order to evaluate the evolution of the wave field, the model is often parabolized or hyperbolized (see [56]), by assuming a progressive wave of the form

$$\eta_{p,l} = a_{p,l} e^{-i\left(-k_l^y y - \int_0^x k_{p,l}^{x'} dx' + \omega_p t\right)}, \qquad (3.10)$$

similar to (3.1) with $k_{p,l}^x$ and k_l^y representing the x- and y-components of the wave number vector respectively. The l-index relates to the discretisation in the lateral direction, where no bottom changes are assumed. This allows the direct satisfaction of the resonance closure in the lateral direction and a decoupling between directional components of each harmonic.

The relation between η_p and $\hat\phi_p$ can be found using a Taylor series expansion of (1.3) about $z = 0$. Note that one should retain $O(\varepsilon^2)$ terms in this relation in order to remain consistent with the equation's order (see [20]). Based on [13] and [73] the deterministic wave evolution equation for the Fourier amplitude of the surface elevation $a_{p,l}$, with constant lateral wavenumber k_l^y, is defined as

$$\frac{1}{C_{g,p}}\frac{\partial a_{p,l}}{\partial t} + \frac{\partial a_{p,l}}{\partial x} + \frac{1}{2C_{g,p}}\frac{\partial C_{g,p}}{\partial x}a_{p,l} + D_{p,l}a_{p,l}$$

$$= -\varepsilon i \sum_{\substack{u=\max\{l-M,-M\}\\s=p-N}}^{\substack{u=\min\{l+M,M\}\\s=N}} W_{s,p-s,u,l-u}a_{s,u}a_{p-s,l-u}e^{-i\int\left(k_{s,u}^x+k_{p-s,l-u}^x-k_{p,l}^x\right)dx}. \tag{3.11}$$

Here, the W-term is the nonlinear interaction kernel defined in Bredmose et al. [13] and Vrecica and Toledo [73] for cases without and with dissipation, respectively. $D_{p,l}$ can describe a linear damping or forcing term, while $t = \varepsilon^2 T$ represents a slow time evolution, which is typically on a different scale than the spatial evolution $x = \varepsilon X$, for T and X physical time and space variables (cf. the comments after (3.2) in Sect. 3.1). It appears when one allows the potential in the mild-slope type equation to vary slowly in time on top of its harmonic behaviour.

Typically wave reflection and nonlinear generation in the backwards direction are second order effects, and are not considered further. They can become significant under certain conditions, and to account for them it is possible either to solve the nonlinear elliptic MSE given in (3.9) or to create two coupled evolution equations— one for forward propagating waves and the other for backward propagating waves in the same manner as in the linear case (see [56]).

3.3 Explanation of Nonlinear Energy Transfer Using Spring-Mass Allegory

One way to think about the wave resonance phenomenon is via an analogy to oscillating mass-spring systems. The linear part of (3.9), upon redefinition of $\hat{\phi}_p$ and k_p, takes the form of a Helmholtz equation (see [56]), which in one dimension becomes a simple harmonic oscillator. Imagine a set of N oscillating spring-mass systems, related to N spectral frequency bins. Softer springs (small spring constant k_p) are in lower harmonics, and as the frequency increases the springs become stiffer (larger k_p). When the problem is linear, these systems are decoupled, but nonlinear terms couple each spring system (spectral bin) to the oscillation of other spring systems.

The nonlinear part of (3.9), which relates to combinations of waves that already satisfy the resonance condition in ω, acts as a forcing term on the mass-spring system, as explained schematically in Sect. 2. These forcing terms are generally small in magnitude, compared to the total energy of the system. Non-resonant forcings (i.e., ones that do not close the resonance condition in k) will cause the system to oscillate slightly at the frequency of forcing (bound wave). If the forcing matches the spring's natural frequency (i.e., the resonance condition in wavenumber is met), a resonance is reached and a significant amount of energy transfers to the related spectral bin.

4 Stochastic Evolution Equations

4.1 Introduction

The deterministic equations given above seem to provide fertile material for modelling the sea. Under the assumption that the waves are not too steep (in particular, not breaking), so as to remain in a weakly nonlinear regime, and that there are no further forces, it seems that if suitable initial conditions can be supplied the subsequent evolution could be solved for numerically. If we are able to measure a sea-surface, and to conclude that it is composed of Fourier modes $\mathbf{k}_i$ with given amplitudes a_i, suitable initial conditions for the discrete Zakharov equation (3.3) consist of specifying $B_i(t = 0)$.

If one is interested in average quantities of the sea-state, like the energy, it becomes necessary to develop new evolution equations, in particular since we cannot accurately specify initial conditions for all situations of interest. In particular, the wave phases are found to be essentially uniformly distributed between $(0, 2\pi]$. Underlying this approach is the idea that the free surface $\eta(x, y, t)$ (or the complex amplitudes in our deterministic equations) is a stochastic process. The perspective taken here is that the temporal evolution of any realization is governed by a given deterministic equation—for example the Zakharov equation (3.2). This is a suitable viewpoint for waves at sea, but in a wave-tank it may be more appropriate to consider a spatial evolution equation instead (see, for example, Shemer and Chernyshova [60]). Our assumption also means that no random forcing by the wind, or the like, plays a role in the evolution of the wave field.

The energy density spectrum, based on linear theory, rests on an underlying assumption of homogeneity of the sea state. This is a prerequisite for sensible measurements (see [31, Sec. 3.5, App. A & App. C], or [40, Sec. 9]) and is a convenient starting point for assumptions that are made in the equations for the temporal evolution of energy spectra. In practical measurements of waves, stationarity (or homogeneity) means that the conditions are unchanged for the duration of the measurement (or over the space being measured). For example, it clearly makes no sense to average two measurements of the sea-surface elevation if one is windward and the other leeward of an island.

While the literature on nonlinear stochastic evolution equations is vast, it is worth pointing the reader to some of the resources with a bearing on water wave theory. The stochastic approach to ocean waves was initiated by Pierson (see [55]) in the 1950s, and an account of the field up to the mid 1960s is found in the engaging work of Kinsman [40]. A general perspective on weakly nonlinear dynamics, also touching on other fields, is provided by Zakharov et al. [79] and Nazarenko

[49], while Janssen [33] places this theory firmly in the context of modern wave forecasting. A particularly clear account of many aspects of nonlinear and random waves may also be found in a book chapter by Trulsen [72].

4.2   Stochastic Evolution Equations for Deep Water

Historically, the first treatment of the evolution of a spectrum of surface waves dates back to Hasselmann [27], shortly after the discovery of resonant interaction theory for surface waves in deep water by Phillips [54]. It is simpler to start with the later work of Longuet-Higgins [44], whose point of departure is the 2D NLS in deep water—as mentioned above, this can be derived from the Zakharov equation.

4.2.1   Narrow-Band Equations

We start, following Longuet-Higgins, with the scaled form of the 2D NLS

$$2i A_\tau = \frac{1}{4}(A_{xx} - 2A_{yy}) + |A|^2 A, \tag{4.1}$$

with $A = A(x, y, \tau)$ the envelope amplitude, x and y slow spatial variables, and $\tau = \varepsilon^2 T$ a slow time. Two approaches are possible, in either physical or in Fourier space, and we explore the former first—the main ideas are identical for both, and can be found, for example, in [79, Sec. 2]

Step 1: write (4.1) at a point $\mathbf{x}_1 = (x_1, y_1)$, and multiply it by $A^*(\mathbf{x}_2) = A^*(x_2, y_2)$, where $*$ stands for a complex conjugate. Step 2: Subtract the equation with $A(\mathbf{x}_1)$ and $A^*(\mathbf{x}_2)$ interchanged. Assume that the envelope amplitudes $A(\mathbf{x}, \tau)$ are stochastic processes, such that each realization is governed by the deterministic NLS (4.1). Step 3: take averages (expected values) of the equation from step 2 to obtain

$$2i \frac{\partial}{\partial \tau}\langle A(\mathbf{x}_1)A^*(\mathbf{x}_2)\rangle = \frac{1}{4}\left(\frac{\partial^2}{\partial x_1^2} - \frac{\partial^2}{\partial x_2^2}\right)\langle A(\mathbf{x}_1)A^*(\mathbf{x}_2)\rangle$$

$$- \frac{1}{2}\left(\frac{\partial^2}{\partial y_1^2} - \frac{\partial^2}{\partial y_2^2}\right)\langle A(\mathbf{x}_1)A^*(\mathbf{x}_2)\rangle \tag{4.2}$$

$$+ \langle A(\mathbf{x}_1)A^*(\mathbf{x}_1)A(\mathbf{x}_1)A^*(\mathbf{x}_2)\rangle - \langle A(\mathbf{x}_2)A^*(\mathbf{x}_2)A(\mathbf{x}_2)A^*(\mathbf{x}_1)\rangle.$$

At this point, further progress depends on stochastic assumptions made for A. The principal obstacle is to treat the fourth-order averages appearing on the right-hand-side of (4.2). Assuming that the process A is close to Gaussian, and has zero mean, allows the decomposition

$$\langle A(\mathbf{x}_1)A^*(\mathbf{x}_1)A(\mathbf{x}_1)A^*(\mathbf{x}_2)\rangle = 2\langle A(\mathbf{x}_1)A^*(\mathbf{x}_1)\rangle\langle A(\mathbf{x}_1)A^*(\mathbf{x}_2)\rangle,$$

where higher order cumulants have been discarded entirely We can factorize the differential operators on the right-hand side of (4.2) by introducing $\mathbf{R} = (r_x, r_y) = \mathbf{x}_1 - \mathbf{x}_2$ and $\mathbf{X} = (X, Y) = \frac{1}{2}(\mathbf{x}_1 + \mathbf{x}_2)$, and with $C(\mathbf{R}, \mathbf{X}) := \langle A(\mathbf{x}_1)A^*(\mathbf{x}_2)\rangle$ rewriting (4.2) as

$$2i\frac{\partial}{\partial\tau}C = \frac{1}{2}\frac{\partial^2}{\partial r_x \partial X}C - \frac{\partial^2}{\partial r_y \partial Y}C + 2C\langle A(\mathbf{x}_1)A^*(\mathbf{x}_1)\rangle - 2C\langle A(\mathbf{x}_2)A^*(\mathbf{x}_2)\rangle.$$

$$(4.3)$$

This is, up to scaling, the deep-water analogue of Alber's equation [5, Eq. (3.7)].

If, in addition, we assume that A is homogeneous in (physical) space, then averages must be invariant under translation, i.e. the autocorrelation C must depend only on spatial separation $\mathbf{R}$, and not on the average position $\mathbf{X}$. Employing this homogeneity condition in (4.3) gives $\partial C/\partial\tau = 0$ at this order, as all terms on the right-hand side of (4.3) vanish. To proceed with a statistically homogeneous theory, the lowest order decomposition of the fourth-order terms $\langle A(\mathbf{x}_1)A^*(\mathbf{x}_1)A(\mathbf{x}_1)A^*(\mathbf{x}_2)\rangle$ must be corrected, by using the product rule and (4.1) in considering $\partial/\partial\tau\langle A(\mathbf{x}_1)A^*(\mathbf{x}_1)A(\mathbf{x}_1)A^*(\mathbf{x}_2)\rangle$. In addition, higher-order cumulants and moments will have to be retained and treated accordingly (see [39]).

Longuet-Higgins [44] pursued exactly such an aim, albeit in Fourier space, substituting

$$A = \sum_n a_n(\tau)e^{i(\lambda_n x + \mu_n y - \omega_n t)}$$

into (4.1), and using the dispersion relation $\omega_n = -\frac{1}{8}(\lambda_n^2 - 2\mu_n^2)$. Equating coefficients, and denoting $\mathbf{k}_i = (\lambda_i, \mu_i)$, he found [44, Eq. 4.3]

$$2i\frac{da_n}{d\tau} = \sum_{p,q,r} a_p a_q a_r^* e^{i(\omega_p + \omega_q - \omega_r - \omega_n)\tau}\delta(\mathbf{k}_p + \mathbf{k}_q - \mathbf{k}_r - \mathbf{k}_n), \quad (4.4)$$

which is formally identical (except for a factor of 2) with (3.3) when the kernel is taken as a constant.[1] To now derive a stochastic evolution equation, follow the three steps above in $\mathbf{k}$−space: multiply (4.4) by a_m^*, subtract the complementary equation, and take averages. Homogeneity in physical space now means a lack of correlation of Fourier modes [53, Eq. (11.75)], so that $\langle a_n a_m^*\rangle = C_n\delta(n - m)$, while the quasi-

---

[1] However, the ω_i satisfy the dispersion relation of the NLS rather than the linear deep-water dispersion relation in (3.3).

Gaussian closure remains unchanged. Finally, this leads to [44, Eq. 4.10], a discrete, narrow-band kinetic equation.

Thus the same deterministic equation has yielded two different stochastic evolution equations, depending on whether or not statistical homogeneity is imposed. Homogeneity means that the energy density (as measured by our correlators) does not change to lowest order, so that the time-scale of evolution is longer. Note in (4.3) that the rate of change of the energy density C is proportional to C, whereas in the homogeneous case [44, Eq. (4.10)] we find $dC/dt \propto C^3$. Thus a homogeneous sea-state can be expected to change only slowly due to nonlinear interactions, except possibly when perturbed by inhomogeneous disturbances.

4.2.2 Stability of Narrow Spectra to Inhomogeneous Disturbances

Alber's equation, which is a finite depth version of (4.3), has proved to be one of the main tools used to study the stability of ocean wave spectra to inhomogeneous disturbances. This is an important question, that has direct bearing on the suitability of modern wave-forecasting codes. To reiterate some main points: an ocean-wave spectrum represents a homogeneous, stationary sea-state, whose energy is transported at the group velocity according to linear theory. Nonlinear wave-wave interaction gives rise to a redistribution of energy from the middle frequencies to lower and higher frequencies, as well as changes in the frequencies (and thus velocities) of the waves themselves [68]. It should be borne in mind that this *homogeneous* energy transfer acts on a timescale of order T/ε^4, which for a typical period T of 10 s, and a typical steepness of $\varepsilon = 0.1$ works out to somewhat more than 27 h.

It is generally appreciated that statistical homogeneity is an idealization— necessary for writing a time-independent spectrum theoretically, and required when measuring waves to establish such spectra at sea (see Hasselmann et al. [29, Sec. 2][2]). In light of this, it is important to establish whether even a small departure from homogeneity might invalidate the conclusions reached based on the homogeneous theory. The question addressed by Alber and others is exactly this: will inhomogeneities give rise to a faster energy exchange, and alter the energy distribution, and thus wave-statistics, of an otherwise homogeneous sea-state.

The case of unidirectional spectra has been particularly well studied, beginning with Alber [5], and recent numerical and analytical work has shed light on many of the central issues. Approaches akin to Alber's, following the linear stability analysis [5, Sec. 4] and arriving at his eigenvalue equation (4.16), have relied on integration (analytic in the case of simple spectral shapes like square, Gaussian, or Lorentz spectra in Stiassnie et al. [66], and numerical for more complex JONSWAP spectra) and parameter studies to establish instability criteria. Simply put, inhomogeneous

---

[2]"Over 2000 wave spectra were measured; about [. . .] 121 corresponded to "ideal" stationary and homogeneous wind conditions." p. 10.

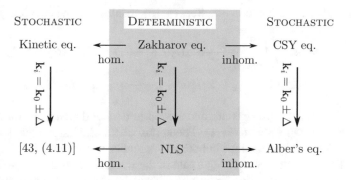

Fig. 2 Diagram of relationships between stochastic and deterministic equations. Indicated are assumptions of statistical homogeneity ("hom.") or inhomogeneity ("inhom."), as well as restrictions to narrow bandwidth ($\mathbf{k}_i = \mathbf{k}_0 \pm \Delta$, $|\Delta| \ll |\mathbf{k}_0|$)

disturbances will grow with the nonlinearity of the wave field, and with decreasing spectral bandwidth; Gramstad [23] has verified that unidirectional JONSWAP spectra are unstable for $\alpha\gamma/\varepsilon > 0.77$, for ε the mean wave slope, and α, γ the usual JONSWAP parameters, using Alber's criteria as well as Monte-Carlo simulations (using the Higher Order Spectral Method). A mathematically rigorous examination of the stability and instability of Alber's equation, including a study of well-posedness, was recently undertaken by Athanassoulis et al. [9] for unidirectional spectra, putting earlier numerical results on solid footing.

For directional sea-states, the matter of instability was investigated by Ribal et al. [58], who were able to extend earlier results for JONSWAP spectra to show that instability also depends on the degree of directional spreading—narrower spectra again being more unstable.

From the deterministic perspective, the Benjamin-Feir instability derived from the NLS is an important mechanism in wave evolution. However, employing the Zakharov equation in place of the NLS [76, Sec. VI.B, Fig. 23ff] yields a more realistic (finite) instability region. The same argument applies to the stochastic counterparts: instability should be investigated not only via the narrow-band NLS, but more generally for the Zakharov equation of which it is a special case (Fig. 2).

4.2.3 Broad-Band Equations

In the above sections, we have discussed stochastic evolution equations derived from the NLS, which implies a narrow bandwidth of order εk_0, for k_0 the carrier wavenumber. As mentioned at the beginning of the section, the equation which models nonlinear interaction in current wave-forecasting codes—Hasselmann's kinetic equation (KE)—has no such restriction. It is possible to derive this equation directly from the Zakharov equation (3.2), using the steps outlined in Sect. 4.2, but retaining terms up to sixth order in the moment hierarchy (see [25, Eq. (2.6)ff]).

Further details can be found in [48, Sec. 14.10], resulting in the equation

$$\frac{dC(\mathbf{k},t)}{dt} = 4\pi \iiint_{-\infty}^{\infty} T_{0,1,2,3}^2 \left(C_2 C_3 (C_0 + C_1) - C_0 C_1 (C_2 + C_3) \right)$$

$$\delta(\mathbf{k} + \mathbf{k}_1 - \mathbf{k}_2 - \mathbf{k}_3)\delta(\omega + \omega_1 - \omega_2 - \omega_3) d\mathbf{k}_1 d\mathbf{k}_2 d\mathbf{k}_3. \tag{4.5}$$

Here, as elsewhere, subscripts are understood to denote dependence on the wavenumber, so that e.g. $C_i = C(\mathbf{k}_i, t)$. The kernel of the Zakharov equation appears again, and due to δ distributions in both wavenumber and frequency it follows that only exactly resonant quartets play a role in the interaction. It is also worth noting that (4.5) predicts no evolution for purely unidirectional waves—symmetric quartets such as $\mathbf{k}_a + \mathbf{k}_b - \mathbf{k}_a - \mathbf{k}_b$ cause the integrand to vanish, and for nonsymmetric unidirectional quartets, the kernel T vanishes [16, p. 147]. This contrasts markedly with the narrow-banded case, where stochastic analogues of the (unidirectional) Benjamin-Feir instability play an important role.

4.2.4 Stability of Broad Spectra to Inhomogeneous Disturbances

A broad-banded evolution equation relaxing the assumption of spatial homogeneity was first derived by Crawford et al. [15], and recently studied for the case of a degenerate quartet of waves by Stuhlmeier and Stiassnie [67]. Like the kinetic equation, it is derived from the Zakharov equation (3.2), and due to the retention of the inhomogeneous terms it has a non-trivial evolution at the same order (and thus, the same time-scale) as the Alber equation (see (4.3)). The discrete version of this equation, which is suitable for numerical computation, is

$$\frac{dr_{nm}}{dt} = i r_{nm}(\omega_m - \omega_n) + 2i \left(\sum_{p,q,r=1}^{N} T_{mpqr} r_{pq} r_{nr} \delta_{mp}^{qr} - \sum_{p,q,r=1}^{N} T_{npqr} r_{qp} r_{rm} \delta_{np}^{qr} \right). \tag{4.6}$$

As mentioned above, underlying the idea of an energy spectrum (for a description of the ocean surface) is the property of statistical homogeneity. The energy spectrum thus consists of the homogeneous terms r_{ii} only, and Eq. (4.6) provides a possibility to study whether such a spectrum undergoes some evolution if suitably perturbed. It is easy to note that if no inhomogeneous terms are present, i.e. the r_{ij} vanish for $i \neq j$, there is no evolution to this order—the next order yields the KE (4.5).

Let us write r_{nm} as $r_{nm} = r_{nm}^h \delta_{nm} + \varepsilon r_{nm}^i$, where the superscripts h and i denote homogeneous and inhomogeneous terms, respectively. Substituting this into (4.6) yields

$$\frac{dr_{nn}^h}{dt} = 0, \tag{4.7}$$

and, for the inhomogeneous terms at order ε:

$$\frac{1}{2i}\frac{dr_{nm}}{dt} = r_{nm}\left(\frac{\omega_m - \omega_n}{2} + \sum_p T_{mppm}r_{pp} - \sum_p T_{nppn}r_{pp}\right)$$

$$+ \sum_{p,q} T_{mpqn}r_{pq}r_{nn}\delta_{qn}^{mp} - \sum_{p,q} T_{npqm}r_{qp}r_{mm}\delta_{qm}^{np}. \quad (4.8)$$

This reduces for a degenerate quartet to the system studied by Stuhlmeier and Stiassnie [67]. Equation (4.8) describes a system of $n^2 - n$ linear, autonomous differential equations. For a given homogeneous initial state, the terms r_{nn} are specified, and the system has the general form

$$\frac{1}{2i}\frac{d\mathbf{r}^i}{dt} = \mathbf{A}\mathbf{r}^i, \quad (4.9)$$

for $\mathbf{A}$ the matrix of coefficients given in (4.8), and $\mathbf{r}^i$ is the vector of the inhomogeneous correlators r_{nm}, $n \neq m$. Negative eigenvalues of $\mathbf{A}$ therefore yield instability. That is, for a given homogeneous state, consisting of a specification of N wave-vectors $\mathbf{k}_1, \ldots, \mathbf{k}_N$ and corresponding energy (or, equivalently, amplitude) in the form $r_{11}, \ldots, r_{NN}$, initially small inhomogeneous disturbances will grow exponentially with time and give rise to energy exchange within the framework of (4.6).

The case of a degenerate quartet $\mathbf{k}_a = (1, 0)$, $\mathbf{k}_b = (1+p, q)$, $\mathbf{k}_c = (1-p, -q)$, which satisfies $2\mathbf{k}_a = \mathbf{k}_b + \mathbf{k}_c$ already demonstrates a range of possible behaviours. Two scenarios are presented in Fig. 3, which depicts the domain of instability (shaded region) for different wave slopes.

For a sea-state with three random waves such that (p, q) is in the shaded region of the figure, the evolution is changed by the presence of small inhomogeneities. One example of this subsequent evolution is given in Fig. 4. The shaded grey region represents a "warm-up" where the inhomogeneous terms (bottom panel) are small, and there is no evolution of the homogeneous terms (top panel). As this is an unstable case, the initially small inhomogeneities grow, and give rise to an energy exchange among the homogeneous terms. Further details on the choice of initial conditions, and the form of the inhomogeneities, may be found in [67].

More realistic cases, involving many modes, for which (4.6) is a generalization of Alber's equation without a narrow-band restriction, await further study.

4.3 Stochastic Evolution Equations for Coastal Environments

While deterministic (often called phase resolving) models can provide relatively accurate calculations of the wave field evolution in coastal waters, they require vast

Fig. 3 Computed region of instability for (4.6) (shaded region), for three waves $\mathbf{k}_a = (1, 0)$, $\mathbf{k}_b = (1 + p, q)$, $\mathbf{k}_c = (1 - p, -q)$, and for different wave slopes. Top panel: $\varepsilon_a = 0.01$, $\varepsilon_b = 0.1$, $\varepsilon_c = 0.1$, the degenerate quartet with greatest growth rate (red dot) has $(p, q) = (0, 0.19)$. Bottom panel: $\varepsilon_a = 0.1$, $\varepsilon_b = 0.01$, $\varepsilon_c = 0.01$, the degenerate quartet with greatest growth rate (red dot) is collinear and has $(p, q) = (0.23, 0)$

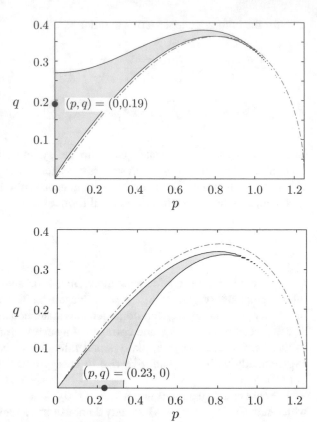

computational resources. This is due to the Nyquist limitation, which enforces small grid spacing in a simulation, and due to a need for a large number of runs required to obtain statistical quantities of interest. Indeed, running nonlinear deterministic models such as the nonlinear MSE (3.9) or high-order Boussinesq (see, e.g., Madsen et al. [46]) for large domains is a very computationally intensive procedure, which commonly reduces the range of practically calculated sea conditions and the size of the modelled region. Extending stochastic models to the nearshore can overcome this restriction by limiting such intensive, deterministic calculations to the very shallow region and the vicinity of coherent marine structures. In addition, it may allow for better nearshore wave forecasting capabilities. Therefore, the extension of stochastic models to the nearshore region is currently of great interest.

4.3.1 One- and Two-Equation Stochastic Models in the Nearshore Region

In the theory of waves in deep-water, cubic nonlinearities give rise naturally to equations for the spectrum in terms of fourth order and sixth order averages (and respective cumulants). Just so, in the nearshore quadratic nonlinearities mean that

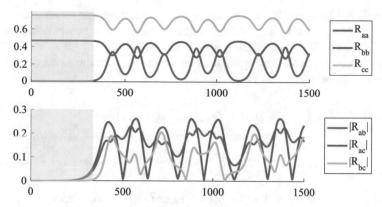

Fig. 4 Depiction of the evolution of a degenerate quartet $\mathbf{k}_a = (1, 0)$, $\mathbf{k}_b = (1.1, 0.2)$, $\mathbf{k}_c = (0.9, -0.2)$ with $\varepsilon_a = 0.01$, $\varepsilon_b = 0.1$, $\varepsilon_c = 0.1$, (see Fig. 3, Top panel) from time $t = 0$ to $t = 1500$ s, under small inhomogeneous disturbances. The top panel depicts the homogeneous terms, while the bottom panel depicts the magnitude of the inhomogeneous terms

the evolution of the spectrum is influenced by third order averages—called the bispectra—as investigated by Hasselmann et al. [28], Elgar and Guza [21] and others. Two-equation nearshore stochastic models consist of an equation for the wave energy evolution (second order moment) with bi-spectral coupling terms, and another equation for the evolution of the bi-spectral components (see [2, 3, 20, 41, 61, 73]). Both equations are commonly derived from the above deterministic models, and have the following general form:

$$
\frac{\partial E_{p,l}}{\partial t_1} + \frac{\partial}{\partial x} \left(C_{g,p} E_{p,l} \right) + 2D_{p,l} C_{g,p} E_{p,l}
$$

$$
= -2C_{g,p} \sum_{\substack{u=\max\{l-M,-M\} \\ s=p-N}}^{\substack{s=N \\ u=\min\{l+M,M\}}} \Re[(i W_{s,p-s,u,l-u} B_{s,u,p-s,l-u}) e^{-i \int (k_{s,u}^x + k_{p-s,l-u}^x - k_{p,l}^x) dx}], \quad (4.10)
$$

which was derived from the deterministic equation (3.11) using the same procedure as in Sect. 4.2.1: multiplying (3.11) by the complex conjugate of $a_{p,l}$, summing the result with its complement, and applying an ensemble average. Here $\Re$ denotes the real part of an expression. The energy spectrum and bispectrum are defined as

$$
E_{p,l} = \left\langle \left| a_{p,l} \right|^2 \right\rangle, \quad B_{s,u,p-s,l-u} = \left\langle a_{p,l}^* a_{s,u} a_{p-s,l-u} \right\rangle. \quad (4.11)
$$

The brackets $\langle \cdot \rangle$ denote the ensemble averaging operation, the index p defines the frequency of the spectral component, l defines its lateral wavenumber, and the terms $D_{p,l}$, $W_{s,p-s,u,l-u}$ are defined as in Sect. 3.2 (see (3.11)).

The bispectrum evolution equation is also derived from the same deterministic model in a similar manner to yield

$$\frac{dB_{s,u,p-s,l-u}}{dx} + \left(D_{p,l} + D_{s,u} + D_{p-s,l-u}\right) B_{s,u,p-s,l-u}$$

$$+ \left(\frac{C'_{g,p}}{2C_{g,p}} + \frac{C'_{g,s}}{2C_{g,s}} + \frac{C'_{g,p-s}}{2C_{g,p-s}}\right) B_{s,u,p-s,l-u} = -i\left(I_{q,r,s-q,u-r,-p,l,p-s,l-u}\right.$$

$$T_{q,r,s-q,u-r,-p,l,p-s,l-u} + I_{q,r,s-q,u-r,-p,l,s,u}T_{q,r,s-q,u-r,-p,l,s,u}$$

$$\left. + I_{q,r,s-q,u-r,s,u,p-s,l-u}T_{q,r,s-q,u-r,s,u,p-s,l-u}\right) \tag{4.12}$$

with the trispectrum components and their coefficients defined as

$$T_{q,r,s-q,u-r,-p,l,p-s,l-u} = \left\langle a_{q,r}a_{s-q,u-r}a^*_{p,l}a_{p-s,l-u}\right\rangle, \tag{4.13}$$

$$I_{q,r,s-q,u-r,-p,l,p-s,l-u} = \sum_{\substack{r=\max\{u-M,-M\}\\q=s-N}}^{\substack{q=N\\r=\min\{u+M,M\}}} W_{q,s-q,r,u-r} e^{-i\int \left(k^x_{q,r}+k^x_{s-q,u-r}-k^x_{s,u}\right)dx}. \tag{4.14}$$

Here, slow time changes of the spectral components were discarded, leading to a formulation of the bispectra as a function of only spatial coordinates.

In a similar manner it can be shown that the trispectrum will depend on fourth order moments, which will in turn depend on fifth order moments and so forth. Therefore, for solving the system a closure relation is required, as when deriving (4.3). A quasi-Gaussian closure [12] is applied to truncate the infinite hierarchy of equations, resulting in

$$\frac{dB_{s,u,p-s,l-u}}{dx} + \left(D_{p,l} + D_{s,u} + D_{p-s,l-u}\right) B_{s,u,p-s,l-u}$$

$$+ \left(\frac{C'_{g,p}}{2C_{g,p}} + \frac{C'_{g,s}}{2C_{g,s}} + \frac{C'_{g,p-s}}{2C_{g,p-s}}\right) B_{s,u,p-s,l-u}$$

$$= -2i\left(W_{-s,-(p-s),u,l-u}E_{s,u}E_{p-s,l-u}\right.$$

$$\left. + W_{p,-s,l,l-u}E_{p,l}E_{s,u} + W_{p,-(p-s),l,l-u}E_{p,l}E_{p-s,l-u}\right) e^{i\int \left(k^x_{s,u}+k^x_{p-s,l-u}-k^x_{p,l}\right)dx}. \tag{4.15}$$

Equations (4.10) and (4.15) comprise a two-equation stochastic model, which can be used to solve the shoaling problem.

The number of permutations between wave components constructing the various bi-spectral components is very large, so that the problem becomes very computationally intensive. In order to address this limitation, Eq. (4.15) is solved for $B_{s,u,p-s,l-u}$. The bispectrum is assumed to be negligible in deep water as in this region the sea is nearly Gaussian. Applying the integrating factor method to Eq. (4.15) yields an analytical solution for the bispectrum, which can then be substituted into (4.10) to construct a one-equation model (see [2, 3, 73]) of the form

$$\frac{\partial E_{p,l}}{\partial t_1} + \frac{\partial}{\partial x}\left(C_{g,p}E_{p,l}\right) + 2D_{p,l}C_{g,p}E_{p,l} \tag{4.16}$$

$$= 4C_{g,p}\sum_{\substack{u=\max\{l-M,-M\}\\s=p-N}}^{\substack{s=N\\u=\min\{l+M,M\}}}\Re\left[Q_{s,p-s,u,l-u} + Q_{p,s,l,t-u} + Q_{p,p-s,l,l-u}\right]W_{s,p-s,u,l-u},$$

$$\tag{4.17}$$

with

$$Q_{s,p-s,u,l-u} = e^{-i\int_0^x\left(k_{s,u}^{x'}+k_{p-s,l-u}^{x'}-k_{p,l}^{x'}\right)dx'}\,e^{-\int_0^x -J_{s,u,p-s,l-u}dx'} \tag{4.18}$$

$$\int_0^x\left(E_{s,u}E_{p-s,l-u}W_{-s,-(p-s),u,l-u}e^{i\int_0^{x'}\left(k_{s,u}^{x''}+k_{p-s,l-u}^{x''}-k_{p,l}^{x''}\right)dx''}\right.$$

$$\left.\times e^{\int_0^{x'} -J_{s,u,p-s,l-u}dx''}\right)dx',$$

where the J-term represents summation of all linear coefficients.

In order to simplify the calculation of the Q-terms, as a first approximation, one can assume slow spectral evolution with respect to shoaling coefficients and take the energy terms outside of the integral (see [2, 3]), similar to the procedure adopted when deriving the kinetic equation (4.5) for the evolution of spectra in deep water (see [25, Eq. (A1)ff]). This may reduce the accuracy in breaking regions where wave heights change significantly within short distances.

4.4 Localization Procedures for Nearshore Stochastic Models

The one-equation model (4.17) reduces the number of equations to be solved significantly. However, the solution still requires the calculation of non-local nonlinear coefficients. This makes its implementation difficult for operational models based on the wave-action equation (WAE). Furthermore, the evaluation of the bispectrum (4.12) or the non-local coefficients (4.18) still enforces a strict Nyquist limitation as they themselves may oscillate quite rapidly in space. A localisation of the Q coefficient can therefore allow for operational model implementation while improving the efficiency of calculation significantly.

Simplified approaches are currently employed in source terms used in operational wave models (see [11, 18, 59]). They approximate the bi-spectral evolution equation using empirical data for the second harmonic, or assume very small changes in the bi-spectra to solve it as an algebraic equation.

A more advanced localization approach, which aims to further improve operational models in the nearshore region, was first considered in Stiassnie and Drimer [62] and subsequently improved in Toledo and Agnon [71] and Vrecica and Toledo [73]. This procedure entails extracting spectral components out of the integral, and only the mean part of the bispectra—the main interest in such models—is accounted for. Assuming a monotone slope, the Q-term is simplified from (4.18) as

$$Q_{s,p-s,u,l-u}(x) = E_{s,u} E_{p-s,l-u} P_{s,p-s,u,l-u}\left(h(x), h'(x)\right), \qquad (4.19)$$

which enables pre-calculation of the nonlinear coupling term P, which is a localized coefficient.

The different behaviours of the non-local nonlinear interaction coefficients (4.18) and the localized simplification (4.19) can be seen in Fig. 5 for a monochromatic wave energy transfer to the second harmonic while shoaling on a beach with a constant slope. In deep water, the nonlinear interaction term (or the bispectra) oscillates in space with no mean change (see Fig. 5 (left panel) in the region of $x < 80$ m). This indicates a bound wave behaviour with mean energy transfer between the modes. Once the wave enters intermediate depths, class III triad resonance conditions (2.3) can be satisfied, and the nonlinear interaction coefficient oscillates, albeit with a small mean change. This indicates a mean energy transfer to the second harmonic in the class III Bragg resonance mechanism as shown in Fig. 1a.

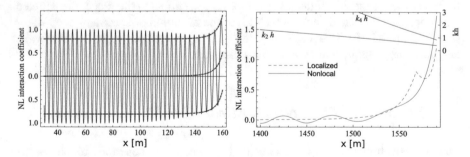

Fig. 5 Two different behaviors of the bispectra for the case of monochromatic wave propagation over a slope, as shown in [73]. In intermediate waters the nonlinear coupling term (Q_{1100}) is mostly oscillating with a slowly growing mean component (left panel). As the water gets shallower (kh reduces) the Q-term tends to an asymptotic shallow water solution (right panel)

As all the interacting waves enter shallow water conditions, they asymptotically go to resonance as in Fig. 1c, so instead of an oscillatory behaviour they act in an exponential manner (see Fig. 5, right panel). Under such conditions, a distinct localization approach is needed compared to the intermediate water case. Such an approach was developed in Vrecica and Toledo [73, Sec. 4.3.2]. The two different formulations of the localized model are separated by a gate term, which is a function of depth and bottom slope (see Fig. 5). It is stressed that the transition between these two behaviours also depends on the bottom slope, for sharp changes the shallow water localization will activate sooner, and vice versa.

A limitation of the model is its inapplicability to long propagation in shallow water with strong nonlinearity. Energy would cascade to ever higher frequencies, leading to wave breaking. This condition is more relaxed for cases with large directional spreading, as waves are still dispersive in the angular space (see [50]). However, operational stochastic models are usually not extended to such areas, typically a Bousinessq-type model (e.g, [46]), or RANS model ([80]) would be used for such cases.

4.5 Comparison Between Deterministic Ensembles and Stochastic Equations

In deep water waves are often considered to be uncorrelated, and the slow evolution of the spectral energy density can be captured by e.g. the kinetic equation (4.5). However, as waves propagate to nearshore correlations build up, and coherent patterns form [51]. Therefore, a quasi-Gaussian closure, while commonly used in nearshore wave models, is often not valid for cases involving strong nonlinearity or dissipation. Such quasi-Gaussian closure can result in an overestimate of energy transfer to higher frequencies, as well as result in (unphysical) negative energies (see [34, Sec. 4.4]). The closure of Holloway [30], which adds dissipation to the bispectral evolution equation, is often used as an empirical solution.

Nonlinear shoaling also affects the wave shape, which is commonly expressed using skewness and asymmetry, as discussed in Elgar and Guza [21]. Initially, in deep water, wave skewness (which relates to wave phase), and asymmetry (which relates to nonlinear energy transfer) are both near zero. As the wave field starts shoaling both begin to grow, however in the surf zone skewness tends back to zero, while the limit for asymmetry is ~3. These depend on the value of the Ursell number $a\lambda^2/h^3$ (for a, λ, h a typical sea-surface elevation, horizontal length scale, and vertical length scale, respectively), as shown in [22].

In order to shed light on the validity of the quasi-Gaussian closure a simulation of unidirectional JONSWAP spectra with a 1 m significant wave height and peak frequency of 0.1 Hz was performed using the nonlinear MSE (3.11). The spectrum propagates from deep water (Fig. 6a) to 5 m depth (Fig. 6b) over a 5% slope. In the present case, the Ursell number is small, so skewness and asymmetry

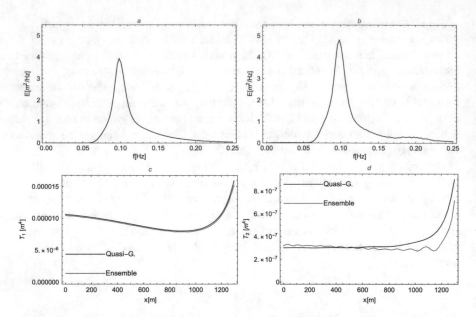

Fig. 6 Evolution of a JONSWAP spectra, obtained using Eq. (3.11), over a monotone (5%) slope from deep water (**a**) to 5 m depth (**b**). Comparison of quasi-Gaussian, and ensemble averaged trispectral moments ($T_1 = T_{39,41,-q,-80+q}$ and $T_2 = T_{-39,80,-q,-41+q}$, lateral indices are dropped for brevity) at 5 m depth, describing energy transfer to secondary peak (**c**), and backtransfer to the primary peak (**d**)

may be neglected. The quasi-Gaussian closure for trispectral moments describing energy transfer to the secondary peak ($E_{39,0}$ and $E_{41,0}$ to $E_{80,0}$), as defined by Eq. (4.11) is compared against ensemble averaged trispectral sums. Comparison is also made between trispectral moments describing backtransfer of energy to $E_{41,0}$. The quasi-Gaussian moments are defined using Eq. (4.13) as $T_1 = E_{39,0}E_{41,0}$ and $T_2 = E_{39,0}E_{80,0}$, while the ensemble averaged ones are defined as $T_1 = \sum_{q=-N}^{N} a_{39,0}a_{41,0}a_q^* a_{80-q,0}^*$, and $T_2 = \sum_{q=-N}^{N} a_{39,0}^* a_{80,0}a_q^* a_{41-q,0}^*$ for the super- and sub-harmonic interactions respectively (N represents the number of discretized wave harmonics). The results are shown in Fig. 6c and d.

As a relatively mild nonlinear case is considered, the quasi-Gaussian closure [12] is accurate to leading order. Based on preliminary analysis, the closure is accurate for the trispectral moments containing the most energetic wave components up to this point. However, the errors are not proportional to each trispectrum, and the relative error is larger for less energetic components. When summed over all possible indices the errors can become significant in cases of strong nonlinearity.

While averaged equations based on the quasi-Gaussian closure may agree well with ensemble averages of the deterministic equations, it is important to point out that individual realizations of the deterministic equations may show significant deviations from the average. This is illustrated via the generation of infragravity waves (0.01 Hz), where Eq. (3.11) is solved with an input of a bichromatic wave

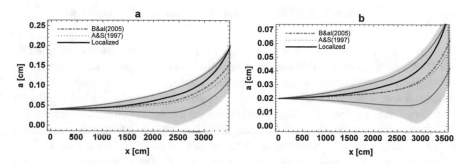

Fig. 7 Evolution of a low-frequency wave (0.01 Hz) of initial amplitudes 0.04 cm (**a**) and 0.02 cm (**b**) due sub-harmonic interaction of bi-chromatic waves (0.11 Hz and 0.1 Hz) over a monotone beach. Grey area and dot-dashed brown line: Monte-Carlo realizations using Eq. (3.11) and their ensemble average. Brown lines: the ensemble's standard deviation added and subtracted from the averaged result see [13]. Thick black line: localized stochastic model. Dotted black line: the ensemble averaged deterministic result of [2]

field (0.1 and 0.11 Hz, with amplitudes of 2.07 cm for Fig. 7a and 1.22 cm for Fig. 7b). The ensemble relates to different relative phases between the bi-chromatic waves. An envelope containing all realizations is shown in grey together with their mean value and standard deviation. It can be clearly seen that depending on the relative phase of interacting components, each realization can be drastically different. Hence, it should be taken into account that the ensemble averaged wave height may be significantly lower than that of the largest realization.

5 Conclusions and Perspectives

We have focused throughout on the mechanism of resonant (or near-resonant) energy exchange, and how it drives wave evolution in deep water as well as finite depth. A variety of deterministic and stochastic model equations exist, suitable for simulating the evolution of a wave field from deep to shallow water. Deterministic models are able to give insight through many realizations, while stochastic models are much faster, enabling analysis of larger areas. Although present day operational wave models are highly reliable overall, there is considerable work to be done on the theoretical front. In particular, while an accuracy within a few percent for a wave forecast may be suitable for the vast majority of situations, it is often the outliers, or extreme events (see, e.g. Adcock and Taylor [1] for a review), that have the most potential to cause damage. We highlight a few areas where further study is needed for deep and shallow water.

5.1 Direct Numerical Simulations and Kinetic Equations

While it is a formally simple step to average an evolution equation, the relationship between solutions of the deterministic equation and its random counterpart is a subtle one, as highlighted above in Sect. 4.5. Ideally, we would observe the following: if averaging is done by assuming, say, phases randomly and independently distributed over $(0, 2\pi]$, then the behaviour of the averaged equation should be the average of many realizations of the deterministic equation, where each realization chooses random phases independently distributed over $(0, 2\pi]$. This is Monte-Carlo simulation, where computationally a "realization" means a solution with given initial data.

The simplest nontrivial case already shows some difficulties: a resonant quartet of four waves. The results of Stiassnie and Shemer [65, Sec. 9, Fig. 8] and Annenkov and Shrira [7, Sec. 3, Fig. 2] demonstrate clear discrepancies. In both cases four initial conditions $b(\mathbf{k}_1, t = 0), \ldots b(\mathbf{k}_4, t = 0)$ are supplied. In the Monte-Carlo simulation, the Zakharov equation is integrated when $b(\mathbf{k}_1, 0) = |b(\mathbf{k}_1, 0)|e^{i\phi}$ for many realizations with different $\phi \in (0, 2\pi]$. The kinetic equation is integrated with the (phase-free) initial condition $C_i(0) = |b(\mathbf{k}_i, 0)|^2$. Annenkov and Shrira [7, Fig. 3] were able to obtain good agreement with the kinetic equation only after replacing each of the four waves in the Monte-Carlo simulation with a cluster of five waves, with a resulting 181 coupled quartets.

The comparison between numerical simulation and the kinetic equation for JONSWAP spectra with many modes has also been extensively explored. Despite the fact that the kinetic equation is derived with some long-time asymptotic limits (to eliminate non-resonant contributions), and is formally on the long time-scale T_4, Tanaka [69] found that it captures spectral changes for a broad $(\cos^2(\theta))$ JONSWAP spectrum on a much shorter time-scale. Tanaka's results point to the fact that ensemble averaging is inessential provided the mode density is high enough—which is related to the theoretical view of the kinetic theory as averaging out redundancy in the dynamic description [79, Sec. 2.1.2]—but the four-wave results point to a lower limit to the applicability of the kinetic theory.

Direct numerical simulations can also be used to investigate other averaged equations. For example, the instability results of Alber [5] were compared with Monte-Carlo simulations of the NLS by Onorato et al. [52, Sec. 3] for unidirectional Lorentz spectra, and some qualitative agreement with the theoretical instability region was found. For initial Gaussian spectra, Dysthe et al. [17] found approximate agreement between Alber's results and Monte-Carlo simulations only in the unidirectional case—for directional Gaussian spectra there were marked discrepancies. All these results, as well as the recent study by Annenkov and Shrira [8] comparing several wave kinetic equations with Monte-Carlo simulations, point to a need for intense further study.

5.2 The Role of Near-Resonant Interactions

The classical kinetic equation (4.5) contains δ-functions in both wavenumber and frequency—this reflects its derivation under the assumptions of exact resonance. This is the consequence of an asymptotic limit, described in detail by Janssen [32, see Eq. 27]. The Zakharov equation requires no such assumption, and includes near-resonant interactions such that $\Delta_{n+p-q-r} = \omega_n + \omega_p - \omega_q - \omega_r$ is of order ε^2. Indeed, the side-band instability relies on exactly these (fast, compared to the kinetic time-scale) interactions.

A number of generalizations of the kinetic equation (4.5) exist, which aim to incorporate near-resonant interactions. Janssen [32, Eq. 25] proposed one such equation, Annenkov and Shrira [7, Eq. 2.25] proposed another, and Gramstad and Stiassnie [25, Eq. 2.19] generalized this to include frequency correction terms. The same generalized equation was recently studied by Andrade et al. [6], and found to exhibit finite-time blow-up for some degenerate quartets. When performing Monte-Carlo simulations, Annenkov and Shrira [7, Sec. 3.4] found that omitting waves in exact resonance, and keeping only their near-resonant neighbors had no effect on the subsequent evolution. There are clearly many more nearly-resonant interactions than there are exactly resonant ones. This fact, together with the need to discretize in wavenumber-space when performing computations, means that some amount of coarse-graining is inevitable. Such near resonant generalizations should be explored in detail, and compared with the kinetic equation (see Annenkov and Shrira [8] for the initiation of such an effort).

5.3 Nearshore Wave Modelling

Much work remains to be done on important aspects of coastal wave modelling. The stochastic nonlinear formulation used in the breaking region (see Sect. 4.5) has limitations, as the quasi-Gaussian closure it employs is not valid in the surf zone. The empirical closure of [30] corrects overestimation of energy transfer, but does not fully describe nearshore wave statistics. Hence, an in-depth study of nearshore wave statistics is still required in order to formulate better stochastic closures. Of particular note is the fact that an arbitrary realization of the deterministic solution can drastically vary from the ensemble averaged result. This can be seen even in the very simplistic case of a subharmonic interaction of bi-chromatic waves (see Fig. 7). For the evolution of a JONSWAP spectrum (Fig. 6), which has numerous phase selection possibilities, is was necessary to generate 200 realizations in order to obtain good agreement between the quasi-Gaussian closure and ensemble averaged trispectra. The work of Smit and Janssen [61] addresses some recent advances in understanding non-Gaussian wave statistics.

Another nearshore phenomenon to consider is wave reflection. Incident waves will be scattered backwards from the bottom slope (see [75] for the linear case) and nonlinearly generated in the backwards direction. In addition, most models assume linear dispersion for every wave harmonic ([36] is good counterexample), and neglect formation of coherent wave patterns, but how important these are is yet to be investigated.

Wave breaking is one of the most important mechanisms in the nearshore region. Its formulations are empirical by nature (usually based on [10]). Therefore, they better resolve cases close to the ones under which they were tested, and commonly require tuning of their coefficients (see [19, 47]). As most laboratory measurements are conducted in wave flumes, the resulting formulations are limited to directly incident waves without addressing two dimensional aspects. Furthermore, when nearshore nonlinear interactions are not well described in the models used for their formulations, these breaking formulations may not separate well between the two mechanisms. Hence, they have problems representing the complex combined behaviour of breaking and nonlinear interaction, and may require changes and recalibration for any advancement in the nonlinear modelling.

Acknowledgements RS is grateful for the hospitality and support of the Erwin Schrödinger Institute for Mathematics and Physics (ESI), Vienna, Austria, as well as support from a Small Grant from the IMA.

References

1. T.A. Adcock, P.H. Taylor, The physics of anomalous ('rogue') ocean waves. Rep. Prog. Phys. **77**, 105901 (2014)
2. Y. Agnon, A. Sheremet, Stochastic nonlinear shoaling of directional spectra. J. Fluid Mech. **345**, 79–99 (1997)
3. Y. Agnon, A. Sheremet, Stochastic evolution models for nonlinear gravity waves over uneven topography. Adv. Coast. Ocean Eng. **6**, 103–133 (2000)
4. Y. Agnon, A. Sheremet, J. Gonsalves, M. Stiassnie, A unidirectional model for shoaling gravity waves. Coast. Eng. **20**, 29–58 (1993)
5. I.E. Alber, The effects of randomness on the stability of two-dimensional surface wavetrains. Proc. Roy. Soc. A **363**, 525–546 (1978)
6. D. Andrade, R. Stuhlmeier, M. Stiassnie, On the generalized kinetic equation for surface gravity waves, blow-up and its restraint. Fluids **4**, 2 (2019)
7. S.Y. Annenkov, V.I. Shrira, Role of non-resonant interactions in the evolution of nonlinear random water wave fields. J. Fluid Mech. **561**, 181–207 (2006)
8. S.Y. Annenkov, V.I. Shrira, Spectral evolution of weakly nonlinear random waves: kinetic description versus direct numerical simulations. J. Fluid Mech. **844**, 766–795 (2018)
9. A.G. Athanassoulis, G.A. Athanassoulis, M. Ptashnyk, T. Sapsis, Landau damping for the Alber equation and observability of unidirectional wave spectra. Preprint. arXiv:1808.05191
10. J.A. Battjes, J.P.F.M. Janssen, Energy loss and set-up due to breaking of random waves, in *Proceedings of the 16th International Conference of Coastal Engineering (ASCE)*, vol. 1 (1978), pp. 569–587
11. F. Becq-Girard, P. Forget, M. Benoit, Non-linear propagation of unidirectional wave fields over varying topography. Coast. Eng. **38**, 91–113 (1999)
12. D.J. Benney, P.G. Saffman, Nonlinear interactions of random waves. Proc. Roy. Soc. London – A **289**, 301–321 (1966)
13. H. Bredmose, Y. Agnon, P.A. Madsen, H.A. Schaffer, Wave transformation models with exact second-order transfer. Eur. J. Mech. B. Fluids **24**(6), 659–682 (2005)

14. P.G. Chamberlain, D. Porter, The modified mild slope equation. J. Fluid Mech. **291**, 393–407 (1995)
15. D.R. Crawford, P.G. Saffman, H.C. Yuen, Evolution of a random inhomogeneous field of nonlinear deep-water gravity waves. Wave Motion **2**(1), 1–16 (1980)
16. A. Dyachenko, V. Zakharov, Is free-surface hydrodynamics an integrable system? Phys. Lett. A **190**(2), 144–148 (1994)
17. K. Dysthe, K. Trulsen, H.E. Krogstad, H. Socquet-Juglard, Evolution of a narrow-band spectrum of random surface gravity waves. J. Fluid Mech. **478**, 1–10 (2003)
18. Y. Eldeberky, J.A. Battjes, Parameterization of triad interactions in wave energy models, in *Coastal Dynamics'95*, ed. by W.R. Dally, R.B. Zeidler (ASCE, Reston, 1995), pp. 140–148
19. Y. Eldeberky, J.A. Battjes, Spectral modelling of wave breaking: application to Boussinesq equations. J. Geophys. Res. **101**, 1253–1264 (1996)
20. Y. Eldeberky, P.A. Madsen, Deterministic and stochastic evolution equations for fully dispersive and weakly nonlinear waves. Coast. Eng. **38**, 1–24 (1999)
21. S. Elgar, R.T. Guza, Observations of bispectra of shoaling of surface gravity waves. J. Fluid Mech. **161**, 425–448 (1985)
22. M.H. Freilich, R.T. Guza, S.L. Elgar, Observations of nonlinear effects in directional spectra of shoaling gravity waves. J. Geophys. Res. Oceans **95**, 9645–9656 (1990)
23. O. Gramstad, Modulational instability in JONSWAP sea states using the alber equation, in *36th International Conference on Ocean, Offshore and Arctic Engineering*, V07BT06A051 (ASME, New York, 2017)
24. O. Gramstad, The Zakharov equation with separate mean flow and mean surface. J. Fluid Mech. **740**, 254–277 (2014)
25. O. Gramstad, M. Stiassnie, Phase-averaged equation for water waves. J. Fluid Mech. **718**, 280–303 (2013)
26. O. Gramstad, Y. Agnon, M. Stiassnie, The localized Zakharov equation: derivation and validation. Eur. J. Mech. B. Fluids **30**, 137–146 (2011)
27. K. Hasselmann, On the non-linear energy transfer in a gravity-wave spectrum Part 1. General theory. J. Fluid Mech. **12**, 481–500 (1962)
28. K. Hasselmann, W. Munk, G. MacDonald, M. Rosenblatt, Time series analysis, in *Bispectra of Ocean Waves* (1963), pp. 125–139
29. K. Hasselmann, T.P. Barnett, E. Bouws, H. Carlson, D.E. Cartwright, K. Enke, J.A. Ewing, H. Gienapp, D.E. Hasselmann, P. Kruseman, A. Meerburg, P. Müller, D.J. Olbers, K. Richter, W. Sell, H. Walden, Measurements of wind-wave growth and swell decay during the Joint North Sea Wave Project (JONSWAP). Technical report, Deutsches Hydrographisches Institut, Hamburg (1973)
30. G. Holloway, Oceanic internal waves are not weak waves. J. Phys. Oceanog. **10**, 906–914 (1980)
31. L. Holthuijsen, *Waves in Oceanic and Coastal Waters* (Cambridge University Press, Cambridge, 2008)
32. P.A.E.M. Janssen, Nonlinear four-wave interactions and freak waves. J. Phys. Oceanog. **33**, 863–884 (2003)
33. P.A.E.M. Janssen, *The Interaction of Ocean Waves and Wind* (Cambridge University Press, Cambridge, 2004)
34. T. Janssen, Nonlinear surface waves over topography. Ph.D. Thesis, University of Delft (2006)
35. P.A.E.M. Janssen, M. Onorato, The intermediate water depth limit of the Zakharov equation and consequences for wave prediction. J. Phys. Oceanog. **37**, 2389–2400 (2007)
36. T.T. Janssen, T.H.C. Herbers, J.A. Battjes, Generalized evolution equations for nonlinear surface gravity waves over two-dimensional topography. J. Fluid Mech. **552**, 393–418 (2006)
37. R.S. Johnson, *A Modern Introduction to the Mathematical Theory of Water Waves* (Cambridge University Press, Cambridge, 1997)
38. J.M. Kaihatu, J.T. Kirby, Nonlinear transformation of waves in finite water depth. Phys. Fluids **8**, 175–188 (1995)

39. Y.C. Kim, E.J. Powers, Digital bispectral analysis and its applications to nonlinear wave interactions. IEEE Trans. Plasma Sci. **7**, 120–131 (1979)
40. B. Kinsman, *Wind Waves* (Dover, New York, 1984)
41. H. Kofoed-Hansen, J.H. Rasmussen, Modeling of nonlinear shoaling based on stochastic evolution equations. Coast. Eng. **33**, 203–232 (1998)
42. V.P. Krasitskii, On reduced equations in the Hamiltonian theory of weakly nonlinear surface waves. J. Fluid Mech. **272**, 1–20 (1994)
43. Y. Liu, D.K.P. Yue, On generalized Bragg scattering of surface waves by bottom ripples. J. Fluid Mech. **356**, 297–326 (1998)
44. M.S. Longuet-Higgins, On the nonlinear transfer of energy in the peak of a gravity-wave spectrum: a simplified model. Proc. R. Soc. A **347**, 311–328 (1976)
45. M.S. Longuet-Higgins, O. M. Phillips, Phase velocity effects in tertiary wave interactions. J. Fluid Mech. **12**, 333–336 (1962)
46. P.A. Madsen, D.R. Fuhrman, B. Wang, A Boussinesq-type method for fully nonlinear waves interacting with a rapidly varying bathymetry. Coast. Eng. **53**, 487–504 (2006)
47. H. Mase, J.T. Kirby, Hybrid frequency-domain KdV equation for random wave transformation, in *Coastal Engineering 1992: Proceedings of the 23rd International Conference* (1992)
48. C.C. Mei, M. Stiassnie, D.K.-P. Yue, *Theory and Applications of Ocean Surface Waves* (World Scientific Publishing Co., Singapore, 2005)
49. S. Nazarenko, *Wave Turbulence*. Lecture Notes in Physics (Springer, Berlin, 2011)
50. A.C. Newell, P.J. Aucoin, Semidispersive wave systems. J. Fluid Mech. **49**, 593–609 (1971)
51. A.C. Newell, B. Rumpf, Wave turbulence. Annu. Rev. Fluid Mech. **43**, 59–78 (2011)
52. M. Onorato, A. Osborne, R. Fedele, M. Serio, Landau damping and coherent structures in narrow-banded $1 + 1$ deep water gravity waves. Phys. Rev. E **67**, 46305 (2003)
53. A. Papoulis, S.U. Pillai, *Probability, Random Variables, and Stochastic Processes*, 4th edn. (McGraw-Hill, New York, 2002)
54. O.M. Phillips, On the dynamics of unsteady gravity waves of finite amplitude Part 1. The elementary interactions. J. Fluid Mech. **9**, 193 (1960)
55. W.J. Pierson, Wind generated gravity waves. Adv. Geophys. **2**, 93–178 (1955)
56. A.C. Radder, On the parabolic equation method for water-wave propagation. J. Fluid Mech. **95**, 159–176 (1979)
57. J.H. Rasmussen, M. Stiassnie, Discretization of Zakharov's equation. Eur. J. Mech. B. Fluids **18**, 353–364 (1999)
58. A. Ribal, A.V. Babanin, I. Young, A. Toffoli, M. Stiassnie, Recurrent solutions of the Alber equation initialized by Joint North Sea Wave Project spectra. J. Fluid Mech. **719**, 314–344 (2013)
59. J. Salmon, P. Smit, T. Janssen, L. Holthuijsen, A consistent collinear triad approximation for operational wave models. Ocean Model. **104**, 203–212 (2016)
60. L. Shemer, A. Chernyshova, Spatial evolution of an initially narrow-banded wave train. J. Ocean Eng. Mar. Energy **3**, 333–351 (2017)
61. P.B. Smit, T.T. Janssen, The evolution of nonlinear wave statistics through a variable medium. J. Phys. Oceanog. **46**, 621–634 (2016)
62. M. Stiassnie, N. Drimer, Prediction of long forcing waves for harbor agitation studies. J. Waterw. Port Coast. Ocean Eng. **132**(3), 166–171 (2006)
63. M. Stiassnie, O. Gramstad, On Zakharov's kernel and the interaction of non-collinear wavetrains in finite water depth. J. Fluid Mech. **639**, 433–442 (2009)
64. M. Stiassnie, L. Shemer, On modification of the Zakharov equation for surface gravity waves. J. Fluid Mech. **143**, 47–67 (1984)
65. M. Stiassnie, L. Shemer, On the interaction of four water waves. Wave Motion **41**, 307–328 (2005)
66. M. Stiassnie, A. Regev, Y. Agnon, Recurrent solutions of Alber's equation for random water-wave fields. J. Fluid Mech. **598**, 245–266 (2008)
67. R. Stuhlmeier, M. Stiassnie, Evolution of statistically inhomogeneous degenerate water wave quartets. Philos. Trans. R. Soc. A **376**, 20170101 (2018)

68. R. Stuhlmeier, M. Stiassnie, Nonlinear dispersion for ocean surface waves. J. Fluid Mech. **859**, 49–58 (2019)
69. M. Tanaka, On the role of resonant interactions in the short-term evolution of deep-water ocean spectra. J. Phys. Oceanog. **37**, 1022–1036 (2007)
70. L.J. Tick, A non-linear random model of gravity waves I. J. Math. Mech. **8**, 643–651 (1959)
71. Y. Toledo, Y. Agnon, Stochastic evolution equations with localized nonlinear shoaling coefficients. Eur. J. Mech. B. Fluids **34**, 13–18 (2012)
72. K. Trulsen, Weakly nonlinear and stochastic properties of ocean wave fields: application to an extreme wave event, in *Waves in Geophysical Fluids*, CISM International Centre for Mechanical Sciences, vol. 489 (Springer, Vienna, 2006), pp. 49–106
73. T. Vrecica, Y. Toledo, Consistent nonlinear stochastic evolution equations for deep to shallow water wave shoaling. J. Fluid Mech. **794**, 310–342 (2016)
74. G.B. Whitham, *Linear and Nonlinear Waves* (John Wiley & Sons, Hoboken, 1974)
75. Y. Yevnin, Y. Toledo, Reflection source term for the wave action equation. Ocean Model. **127**, 40–45 (2018)
76. H.C. Yuen, B.M. Lake, Nonlinear dynamics of deep-water gravity waves, in *Advances in Applied Mechanics* (Academic Press, Cambridge, 1982), pp. 68–229
77. V. Zakharov, Stability of periodic waves of finite amplitude on the surface of a deep fluid. J. Appl. Mech. Tech. Phys. **9**, 190–194 (1968)
78. V.E. Zakharov, Inverse and direct cascade in a wind-driven surface wave turbulence and wave-breaking, in *IUTAM Symposium*, Sydney, ed. by M.L. Banner, R.H.J. Grimshaw (Springer, Berlin, 1992), pp. 69–91
79. V.E. Zakharov, V.S. L'vov, G. Falkovich, *Kolmogorov Spectra of Turbulence I* (Springer, Berlin, 1992)
80. M. Zijlema, G. Stelling, P.B. Smit, SWASH: An operational public domain code for simulating wave fields and rapidly varied flows in coastal waters. Coast. Eng. **58**, 992–1012 (2012)

Gravity-Capillary and Flexural-Gravity Solitary Waves

Emilian I. Părău and Jean-Marc Vanden-Broeck

Abstract Solitary gravity-capillary and flexural-gravity waves in two and three dimensions of space are reviewed in this paper. Numerical methods used to compute the solitary waves are described in detail and typical solutions found over the years are presented. Similarities and differences between the solutions for the two physical problems are discussed.

Keywords Solitary waves · Flexural-gravity · Gravity-capillary

Mathematics Subject Classification (2000) Primary 76B25; Secondary 76B45

1 Introduction

Solitary gravity-capillary waves at the surface a inviscid fluid have been investigated intensively for the last 40 years (see [17, 65] for reviews). Due to the similarities with the gravity-capillary problem and motivated by the observation of waves under continuous ice sheets, the interest in flexural-gravity solitary waves has also increased considerably in the last 15–20 years.

Under certain conditions the floating ice plates can be modelled using the theory of elastic plates or shells (see [46, 62]). Different models for the elastic plates have been proposed over the years, starting with linear plates [29], Kirchhoff–Love plates [22, 23] or, more recently, using the special Cosserat theory of hyperelastic shells satisfying Kirchhoff's hypothesis [61].

In this paper we will review the most utilised numerical methods in the computation of solitary gravity-capillary and flexural-gravity waves in two and three

E. I. Părău (✉)
School of Mathematics, University of East Anglia, Norwich, UK
e-mail: e.parau@uea.ac.uk

J.-M. Vanden-Broeck
Department of Mathematics, University College London, London, UK
e-mail: j.vanden-broeck@ucl.ac.uk

© Springer Nature Switzerland AG 2019
D. Henry et al. (eds.), *Nonlinear Water Waves*, Tutorials, Schools, and Workshops
in the Mathematical Sciences, https://doi.org/10.1007/978-3-030-33536-6_11

dimensions. Different types of solitary waves will be discussed for both physical problems. While we concentrate here on the numerical solutions of the fully-nonlinear equations of motion, it is worth mentioning that there are also numerous results obtained for various weakly-nonlinear model equations and rigorous proofs of existence of gravity-capillary or flexural-gravity solitary in two and three dimensions.

When a shallow water waves approximation is used weakly-nonlinear model equations of the KdV-type [47] in two dimensions have been generalised to three-dimensions, and KP-type [43] equations were obtained which admit fully-localised solitary waves solutions in the gravity-capillary case [2, 8, 18, 50] (see also [4] for a review of KP solutions). Higher order weakly-nonlinear model equations such as fifth-order KdV which have solitary wave packets as solutions have also been derived for some critical region of parameters for gravity-capillary waves [30, 74] or flexural-gravity waves [34, 73] in two dimensions, and higher order KP-type equations were derived for the three-dimensional flexural-gravity case [36, 37].

Weakly-nonlinear model equations which admit packet-type solitary waves solutions have also been derived by removing the shallow water waves assumption. In two dimensions they are of the NLS-type (see e.g [3, 15, 49]) and in three-dimensions they are of the Benney–Roskes–Davey–Stewartson (BRDS)-type, derived initially for gravity waves [7, 14] and latter generalised for gravity-capillary waves [20]. More recent studies of these weakly-nonlinear equations have been conducted for gravity-capillary waves [1, 5, 44] and for flexural-gravity waves [6, 54].

The existence of gravity-capillary solitary waves in two dimensions has been proved in two dimensions using a spatial dynamics method and centre-manifold techniques in the strong surface tension case [45] and in the weak surface tension case [40–42]. The existence of flexural-gravity solitary waves in two dimensions was established using variational method for the Cosserat model [32] (see also [16, 39] for results using the Kirchhoff–Love model). Recently, the existence of three-dimensional fully-localised gravity-capillary solitary waves was also proved using variational methods for strong tension [9, 31] and weak tension [10].

In this paper in Sect. 2 the two physical problems are formulated. In Sect. 3 two-dimensional solitary waves are presented, including some popular numerical methods used to compute them. In Sect. 4 the three-dimensional solitary waves are discussed briefly, together with a short description of a boundary-integral equation method and the paper ends with conclusions in Sect. 5.

2 Formulation

We consider an inviscid, incompressible fluid of constant density ρ and an irrotational flow in Cartesian coordinates $Oxyz$, with z being the vertical coordinate. We assume that the fluid is bounded below by a rigid bottom at $z = -h$. The free surface or ice/water interface is given by $z = \zeta(x, y, t)$ and we introduce the velocity potential $\Phi(x, y, z, t)$ in the fluid. The governing equation in the fluid domain is

$$\nabla^2 \Phi = 0 \qquad \text{for } x, y \in \mathbf{R}, -h < z < \zeta(x, y, t). \qquad (2.1)$$

The kinematic condition on the free surface is

$$\zeta_t + \Phi_x \zeta_x + \Phi_y \zeta_y = \Phi_z \qquad \text{on } z = \zeta(x, y, t), \qquad (2.2)$$

and the no-flow condition at the bottom is

$$\Phi_z = 0 \qquad \text{on } z = -h. \qquad (2.3)$$

In the deep-water case the last condition is replaced by

$$|\nabla \Phi| \to 0 \qquad \text{as } z \Longrightarrow -\infty. \qquad (2.4)$$

The dynamic boundary condition on the free surface is

$$\Phi_t + \frac{1}{2} \left(\Phi_x^2 + \Phi_y^2 + \Phi_z^2 \right) + g\zeta + P = 0 \text{ on } z = \zeta(x, y, t), \qquad (2.5)$$

where g is the gravitational acceleration. When considering only the effect of the surface tension σ we replace P by $P = \frac{\sigma}{\rho} P_{gc}$, where

$$P_{gc} = - \left[\left(\frac{\zeta_x}{\sqrt{1 + \zeta_x^2 + \zeta_y^2}} \right)_x + \left(\frac{\zeta_y}{\sqrt{1 + \zeta_x^2 + \zeta_y^2}} \right)_y \right]. \qquad (2.6)$$

If we assume that the fluid is covered instead by an ice sheet modelled using the Cosserat theory of hyperelastic shells [61], then the dynamic boundary condition at the interface between the fluid and the ice sheet $z = \zeta(x, y, t)$ is still (2.5), but now $P = \frac{D}{\rho} P_{fg}$, describing the effect of the ice on the surface of water, where D is the flexural rigidity of the elastic shell and

$$P_{fg} = \frac{2}{\sqrt{a}} \left[\partial_x \left(\frac{1 + \zeta_y^2}{\sqrt{a}} \partial_x H \right) - \partial_x \left(\frac{\zeta_x \zeta_y}{\sqrt{a}} \partial_y H \right) - \partial_y \left(\frac{\zeta_x \zeta_y}{\sqrt{a}} \partial_x H \right) + \partial_y \left(\frac{1 + \zeta_x^2}{\sqrt{a}} \partial_y H \right) \right]$$
$$+ 4H^3 - 4KH \qquad (2.7)$$

where H is the mean curvature and K the Gauss curvature of the ice/water interface, given by

$$H = \frac{1}{2a^{3/2}}\left[(1+\zeta_y^2)\zeta_{xx} - 2\zeta_{xy}\zeta_x\zeta_y + (1+\zeta_x^2)\zeta_{yy}\right]$$

$$K = \frac{1}{a^2}\left[\zeta_{xx}\zeta_{yy} - \zeta_{xy}^2\right]$$

$$a = 1 + \zeta_x^2 + \zeta_y^2.$$

We assume here that the ice sheet is not pre-stressed and there is no friction between ice sheet and the fluid [35]. The effect of inertia of the plate is also neglected. We also note that $P_{gc} = -2H$. If we consider a linear elastic plate [57], the P_{fg} term simplifies to the bilaplacian term

$$P_{fg} = \zeta_{xxxx} - 2\zeta_{xxyy} + \zeta_{yyyy}.$$

By looking for linear waves of the form $e^{i(kx+ly-\omega t)}$, we can derive the dispersion relation in the two cases. For the surface tension case it is

$$\omega^2 = \left(g|\mathbf{k}| + \frac{\sigma}{\rho}|\mathbf{k}|^3\right)\tanh(|\mathbf{k}|h), \tag{2.8}$$

and for the ice-covered fluid

$$\omega^2 = \left(g|\mathbf{k}| + \frac{D}{\rho}|\mathbf{k}|^5\right)\tanh(|\mathbf{k}|h), \tag{2.9}$$

where $\mathbf{k} = (k, l)$, hence $|\mathbf{k}| = \sqrt{k^2 + l^2}$.

If we investigate two-dimensional waves which are travelling in x-direction and are constant on the transverse direction, and assume now that y is the vertical coordinate instead of z, the problem will simplify: the free surface or ice/water interface is given by $y = \eta(x, t)$, the velocity potential in the fluid is $\Phi(x, y, t)$ and they will satisfy the equivalent two-dimensional equations corresponding to (2.1)–(2.5). It is worth writing explicitly only the dynamic boundary condition.

$$\Phi_t + \frac{1}{2}\left(\Phi_x^2 + \Phi_y^2\right) + g\eta + P = 0, \quad \text{on } y = \eta(x, t). \tag{2.10}$$

For the surface tension case

$$P = \frac{\sigma}{\rho}P_{gc} = -\frac{\sigma}{\rho}\kappa, \quad \text{on } y = \eta(x, t), \tag{2.11}$$

where κ is the curvature of the free surface given by

$$\kappa = \frac{\eta_{xx}}{(1 + \eta_x^2)^{3/2}}.$$

For the ice-covered fluid case

$$P = \frac{D}{\rho} P_{fg} = \frac{D}{\rho} \left(\kappa_{ss} + \frac{1}{2}\kappa^3 \right), \quad \text{on } y = \eta(x, t), \tag{2.12}$$

where s is the arclength along this interface, and therefore

$$\kappa_{ss} + \frac{1}{2}\kappa^3 = \frac{1}{\sqrt{1 + \eta_x^2}} \partial_x \left[\frac{1}{\sqrt{1 + \eta_x^2}} \partial_x \left(\frac{\eta_{xx}}{(1 + \eta_x^2)^{3/2}} \right) \right] + \frac{1}{2} \left(\frac{\eta_{xx}}{(1 + \eta_x^2)^{3/2}} \right)^3.$$

Simplified Kirchhoff–Love leading-order versions of P_{fg} have been used in the past as approximations of the elastic plate [55, 67], e.g.

$$P_{fg} = P_{KL} = \kappa_{xx} = \partial_{xx}^2 \left(\frac{\eta_{xx}}{(1 + \eta_x^2)^{3/2}} \right). \tag{2.13}$$

To simplify the analysis, we will non-dimensionalise all the equations and variables, using in the gravity-capillary case the length and times units $L_{gc} = (\sigma/\rho g)^{1/2}$ and $T_{gc} = (\sigma/\rho g^3)^{1/4}$, and in the flexural-gravity case $L_{fg} = (D/\rho g)^{1/4}$ and $T_{fg} = (D/\rho g^5)^{1/8}$. We also introduce the dimensionless depths $h_{gc} = h/L_{gc} = h(\rho g/\sigma)^{1/2}$ in the gravity-capillary case and $h_{fg} = h/L_{fg} = h(\rho g/D)^{1/4}$ in the flexural-gravity case. The dimensionless dispersion relations for two-dimensional gravity-capillary waves moving in the x-direction with wavenumber k with phase-speed $c = \omega/k$ is

$$c = \sqrt{\left(\frac{1}{k} + k \right) \tanh \left(kh_{gc} \right)}, \tag{2.14}$$

and in the ice-covered fluid case

$$c = \sqrt{\left(\frac{1}{k} + k^3 \right) \tanh \left(kh_{fg} \right)}. \tag{2.15}$$

It can be observed that in both cases $c \to \infty$ as $k \to \infty$. The long-wave limit $k \to 0$ of $c(k)$ is $c_0 = \sqrt{h_{gc}}$ for gravity-capillary waves and $c_0 = \sqrt{h_{fg}}$ for flexural-gravity waves.

When the water is of finite depth there is always a minimum of the phase speed $c = c(k)$ in the flexural-gravity case, while in the gravity-capillary case this minimum exists only when $\sigma/\rho g h^2 < 1/3$ (see Fig. 1). This can be easily seen if

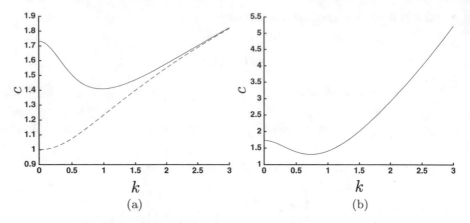

Fig. 1 Dispersion relation for (**a**) gravity-capillary waves given by (2.14) and (**b**) flexural-gravity waves given by (2.15). The dimensionless parameters are $h_{gc} = 3$ (solid line) and $h_{gc} = 1$ (dashed line) for gravity-capillary waves in (**a**) and $h_{fg} = 3$ and flexural-gravity waves in (**b**)

we expand (2.14) in Taylor series about $k = 0$ for gravity-capillary waves

$$c \approx \sqrt{h_{gc}} + \frac{(3 - h_{gc}^2)\sqrt{h_{gc}}}{6}k^2 + O(k^4),$$

and expand (2.15) about $k = 0$ for flexural-gravity waves

$$c \approx \sqrt{h_{fg}} - \frac{h_{fg}^{5/2}}{6}k^2 + O(k^4).$$

If the water is of infinite depth, then $c \to \infty$ as $k \to 0$ and there is always a minimum phase speed $c_{\min} = \sqrt{2} \approx 1.4142$ at $k_{\min} = 1$ for gravity-capillary waves and $c_{\min} = \frac{2}{3^{3/8}} \approx 1.3247$ at $k_{\min} = 3^{1/4}$ for flexural-gravity waves.

Different types of solitary waves can bifurcate from either the long-wave limit speed c_0, or from the minimum phase speed $c_{\min}$ (when it exists).

3 Two-Dimensional Solitary Waves

3.1 Numerical Methods

We present some numerical methods used to compute solitary waves. We concentrate only on computations of steady gravity-capillary and flexural-gravity waves here. One of the methods used to calculate solitary waves is based on boundary integral equation technique [66].

For steady waves in a reference frame moving at constant speed c, we introduce the complex velocity potential

$$w(z) = \Phi(x, y) + i\,\Psi(x, y), \tag{3.1}$$

where $\Psi(x, y)$ is the stream function and $x = x - ct$. We map the physical plane

$$z = x(w) + i\,y(w),$$

to $w(z)$ in the inverse plane. Without loss of generality, we set $\Psi = 0$ on the free surface or fluid-ice interface and choose $\Phi = 0$ at $x = 0$. It can be shown that $\Psi = -ch$ on the bottom (here $h = h_{gc}$ or $h = h_{fg}$, depending on the problem). The fluid-ice interface is denoted by

$$(x(\Phi), y(\Phi)) = (x(\Phi + i\,0), y(\Phi + i\,0)).$$

In this notation, $x'(\Phi)$ and $y'(\Phi)$ are the values of x_Φ and y_Φ evaluated at the interface $\Psi = 0$. As $y_\Phi = 0$ on the bottom $\Psi = -ch$ for the finite-depth problem, we can extend the function $x_\Phi - 1/c + iy_\Phi$ by symmetry about the line $\Psi = -ch$ to an analytic function in the strip $(-2ch, 0)$, then apply the Cauchy integral formula in this rectangular strip.

Assuming the symmetry of solutions about $\Phi = 0$, application of the Cauchy integral formula yields, after some algebra,

$$x'(\Phi_0) - \frac{1}{c} = -\frac{1}{\pi}\!\int_0^\infty y'(\Phi)\left(\frac{1}{\Phi - \Phi_0} + \frac{1}{\Phi + \Phi_0}\right)d\Phi$$

$$+ \frac{1}{\pi}\int_0^\infty \frac{(\Phi_0 - \Phi)y'(\Phi) + 2ch(x'(\Phi) - 1/c)}{(\Phi - \Phi_0)^2 + 4c^2h^2}\,d\Phi$$

$$+ \frac{1}{\pi}\int_0^\infty \frac{-(\Phi_0 + \Phi)y'(\Phi) + 2ch(x'(\Phi) - 1/c)}{(\Phi + \Phi_0)^2 + 4c^2h^2}\,d\Phi, \tag{3.2}$$

where the primes denote differentiation with respect to Φ and the evaluation point Φ_0 lies on the free surface. In the infinite-depth case only the integral on the first line remains, which is evaluated in the principal value sense [66].

The dimensionless dynamic condition becomes in the inverse plane in the gravity-capillary case

$$\frac{1}{2}\left(\frac{1}{x'^2 + y'^2} - c^2\right) + y - \frac{y''x' - y'x''}{(x'^2 + y'^2)^{3/2}} = 0, \tag{3.3}$$

and in the flexural-gravity case

$$\frac{1}{2}\left(\frac{1}{x'^2 + y'^2} - c^2\right) + y + \frac{1}{2}\left(\frac{y''x' - y'x''}{(x'^2 + y'^2)^{3/2}}\right)^3 + \frac{S}{(x'^2 + y'^2)^{9/2}} = 0, \qquad (3.4)$$

where

$$
\begin{aligned}
S = &\; x'^5 y^{(iv)} + 2x'^3 y'^2 y^{(iv)} + x' y'^4 y^{(iv)} - 6x'^4 x'' y''' - 2x'^2 y'^2 x'' y''' \\
&+ 4x'' y'^4 y''' - x'^4 x^{(iv)} y' - 2x'^2 x^{(iv)} y'^3 - x^{(iv)} y'^5 - 4x'^4 x''' y'' \\
&+ 2x'^2 x''' y'^2 y'' + 6x''' y'^4 y'' - 10x'^3 y' y'' y''' + 10x'^3 x'' x''' y' \\
&+ 10x' x'' x''' y'^3 - 10x' y'^3 y'' y''' - 39x' x''^2 y'^2 y'' + 3x''^3 y'^3 - 3x'^3 y''^3 \\
&+ 15x' y'^2 y''^3 + 39x'^2 x'' y' y''^2 - 15x'' y'^3 y''^2 + 15x'^3 x''^2 y'' - 15x'^2 x''^3 y'.
\end{aligned}
$$

Equations (3.2) and (3.3) or (3.4) define a system for the unknown functions $x(\Phi)$ and $y(\Phi)$ which is solved by employing a method described in [33, 34, 66]. The system is discretized by choosing n equally spaced points $\Phi_j = j\Delta\Phi$ for $j = 1, \ldots, n$. The integral (3.2) is evaluated at mid-points by the trapezoidal rule. Finite differences and interpolation formulae are used for the derivatives. Equation (3.3) or (3.4) is evaluated at the interior grid points, and a truncation condition at Φ_n is imposed. The nonlinear system obtained for the unknowns $y'_i = y'(\Phi_i)$, where $i = 1, \ldots, n$, is solved by Newton's method.

A different method based on conformal mapping methods is also used to compute gravity-capillary or flexural-gravity solitary waves [21]. As above, the main idea is to reformulate the physical system which involves an unknown free surface as a system on a fixed domain in a new complex plane $\xi + i\beta$, using a conformal map. In the transformed plane the free surface corresponds to $\beta = 0$ and it is described parametrically as $(X(\xi), Y(\xi))$. The link between these variables is in infinite-depth

$$X(\xi) = \xi - \mathcal{H}[Y(\xi)], \qquad (3.5)$$

where $\mathcal{H}$ is the Hilbert transform

$$\mathcal{H}[f(\xi)] = \int_{-\infty}^{\infty} \frac{f(\xi')}{\xi' - \xi} d\xi'.$$

In the finite case the physical domain is mapped into a strip of depth $\bar{h}$ in the transformed plane and the operator $\mathcal{H}$ in (3.5) changes to $\mathcal{T}$, where

$$\mathcal{T}[f(\xi)] = \frac{1}{2\bar{h}} \int_{-\infty}^{\infty} f(\xi') \coth\left(\frac{\pi}{2\bar{h}}(\xi' - \xi)\right) d\xi'.$$

The dynamic condition to be solved in the steady case of a gravity-capillary wave propagating with speed c is

$$\frac{c^2}{2}\left(\frac{1}{J} - 1\right) + Y - \frac{Y_{\xi\xi}X_\xi - Y_\xi X_{\xi\xi}}{J^{3/2}} = 0, \tag{3.6}$$

where $J = X_\xi^2 + Y_\xi^2$ is the Jacobian of the conformal map. A similar equation can be easily written for flexural-gravity waves, with the last term replaced by terms as in (3.4) and the primes replaced by ξ-derivatives.

The solitary waves are approximated by long periodic waves and the solutions are written in Fourier series with unknown coefficients. The Fourier series for Y is truncated after N terms and the same number of collocation points are distributed along the ξ-axis. A set of algebraic equations is obtained from Eq. (3.6), or a similar one for flexural-gravity waves, which is then solved using the Newton method.

The generalised solitary waves, which are waves with a central pulse and non-decaying oscillations, can be also approximated with long periodic waves and can be computed with this numerical method based on conformal mapping. They were also computed using a modification of the method based on Cauchy integral methods described above, where the integrals in (3.2) will contain cotangent terms [38]. Another approach to compute generalised solitary waves is to use a truncation series method, as described in detail in [65].

3.2 Numerical Results

We will describe the main results obtained on the two problems under consideration, highlighting the differences between them.

In water of infinite depth there exists branches of symmetric solitary waves of elevation and depression which bifurcate from $c = c_{min}$ and continue for $c < c_{min}$ in both flexural-gravity and gravity-capillary cases [48, 66]. They are characterised by damped oscillations in the direction of propagation (see Fig. 2 for an example) The main difference between the two cases is that the branches of gravity-capillary solitary waves bifurcate from zero-amplitude periodic solutions at $c = c_{min}$, while the flexural-gravity solitary waves bifurcate from a finite-amplitude solution at $c = c_{min}$. The solitary waves end in a trapped-bubble for the depression branch [33, 66]. The flexural-gravity waves branch can be continued to $c = 0$, where a self-intersecting profile is obtained, but the solutions are obviously not physical past the trapped-bubble point. The elevation branch is more complicated in both cases, having a number of turning points (see [19] for gravity-capillary waves and [70], Fig. 3 for flexural-gravity waves).

More recently a plethora of non-symmetric and multi-hump gravity-capillary [72] and flexural-gravity waves [26] have been discovered and computed (see Fig. 3 (left) for an example)). The initial guess for the numerical scheme to find

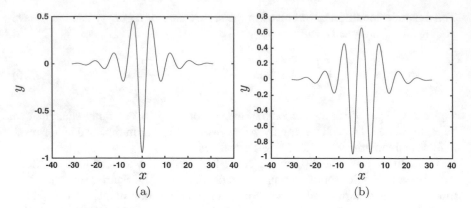

Fig. 2 Symmetric flexural-gravity solitary waves in infinite depth: depression (**a**, left) and elevation solitary waves (**b**, right). In both cases $c = 1.29$

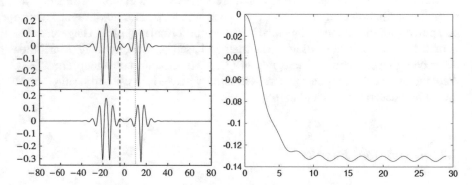

Fig. 3 Two non-symmetric gravity-capillary wave in infinite depth with $c = 1.385$ (left). Symmetric generalised gravity-capillary solitary wave (right). Only half of the wave is shown

these solutions was obtained by 'glueing' different elevation or depression waves travelling at the same speed.

Finite depth has an important effect on the branches of solutions in both cases. Elevation and depression solitary waves bifurcating from c_{min} (when exist) were found. When the depth of the fluid is less than a critical one the branches of flexural-gravity solitary waves bifurcate from zero-amplitude solutions at $c = c_{min}$, as do the gravity-capillary solitary waves.

In the gravity-capillary case, when $\sigma/g\rho h^2 > 1/3$ a different type of depression waves exists when $c < c_0$, with the branch of solutions bifurcating from c_0. For small amplitudes they are very similar with $sech^2$ solutions of KdV equation and they decay monotonically to infinity. There are no equivalent solutions in the flexural-gravity case.

However, when $\sigma/g\rho h^2 < 1/3$ for the gravity-capillary case and for any values of parameters for flexural-gravity case, there are branches of generalised solitary waves bifurcating from c_0 and they exist for $c > c_0$. These are waves with one or more central pulses with non-decaying oscillations. Symmetric single-hump gravity-capillary generalised solitary waves have been computed in [11, 38, 64], where the central pulse is always of elevation (see Fig. 3 for an example). More recently symmetric flexural-gravity generalised solitary waves were computed in [67] for a Kirchhoff-Love model (2.13) and in [25] for the general elastic model (2.12). Multi-humped symmetric generalised solitary waves have also been computed using conformal mapping techniques and pseudospectral methods to solve the fully-nonlinear gravity-capillary waves problem rewritten as a Babenko-type equation [12].

Using the conformal mapping techniques described above (3.5) and (3.6) new branches of non-symmetric generalised solitary waves have been discovered in the last few years: gravity-capillary waves in [27] and flexural-gravity waves in [28].

It is worth noting that generalised flexural-gravity solitary waves have also been found in infinite depth for $c > c_{min}$, when the solitary waves with decaying oscillations existing for $c < c_{min}$ become generalised solitary waves at $c = c_{min}$ (see [52, 53]).

The evolution in time and the stability of the computed solitary waves were studied using numerical methods based on time-dependent conformal maps [51, 52] or by using a Hamiltonian reformulation of the problem and the truncation of Dirichlet-to-Neumann operator [13, 33, 34] which gives approximations of the normal velocity at the free surface. Both methods use pseudospectral techniques based on the fast Fourier transform.

4 Three-Dimensional Solitary Waves

4.1 Numerical Methods

The numerical computation of solitary waves in three-dimensions can be performed using boundary integral equations methods based on Green's theorem [24, 56]. After some manipulation of the Laplace equation (2.1) and using Green second identity we obtain

$$\frac{1}{2}(\Phi(P^*) - x^*) = \int \int_S \left[(\Phi(P) - x) \frac{\partial G(P, P^*)}{\partial n} - G(P, P^*) \frac{\partial (\Phi(P) - x)}{\partial n} \right] dS,$$
(4.1)

where n is the normal to the free surface or ice-water interface S pointing into the fluid, and P^* is a point from S. The Green's function in infinite depth for the points $P = (x, y, z)$ and $P^* = (x^*, y^*, z^*)$ is

$$G(P, P^*) = \frac{1}{4\pi} \frac{1}{((x - x^*)^2 + (y - y^*) + (z - z^*)^2)^{(1/2)}},$$
(4.2)

The Green function is modified when the water is of finite depth is considered by including a term taking into account the symmetry on the bottom.

By defining

$$\phi(x, y) = \Phi(x, y, \zeta(x, y)), \tag{4.3}$$

it allows us to rewrite (4.1) in terms of surface integrals. The dimensionless dynamic boundary condition is

$$\frac{1}{2} \frac{(1 + \zeta_x^2)\phi_y^2 + (1 + \zeta_y^2)\phi_x^2 - 2\zeta_x\zeta_y\phi_x\phi_y}{1 + \zeta_x^2 + \zeta_y^2} + \zeta + P = \frac{c^2}{2} \tag{4.4}$$

here $P = P_{gc}$ or P_{fg}, depending on the problem studied (see [59, 60] for gravity-capillary waves, [57, 58, 63] for flexural-gravity waves). Equations (4.1) and (4.4) are discretised by setting x_i and y_j to be equally spaced points such that $i = 1, \ldots, N$ and $j = 1, \ldots, M$ and the resulting algebraic equations are solved using Newton's method.

Three-dimensional solitary waves have also been computed by numerically solving model equations obtained by reformulating the equations of motion in a Hamiltonian form and then truncating the Dirichlet-to-Neumann operator associated up to some order (see [68] for gravity-capillary solitary waves and [71] for flexural-gravity waves).

4.2 Numerical Results

With the methods described above fully-localised solitary waves have been computed for the two problems on both infinite and finite depth. As for the two-dimensional case, branches of elevation and depression solitary waves have been found bifurcating from the c_{min} for all values of parameters in the flexural-gravity case and when $\sigma/\rho g h^2 < 1/3$ in the gravity-capillary case. These waves have a central depression or elevation and have decaying oscillations in the direction of propagation, but decay monotonically in the transverse direction (see Figs. 4 and 5 for some examples).

In the gravity-capillary case the amplitude of the solitary waves decays to zero as $c \nearrow c_{min}$, approaching a train of two-dimensional periodic waves of zero-amplitude for both infinite and finite depth [59, 60]. In the flexural-gravity case a similar behaviour is found in finite depth when h is small. However, when h is larger or infinite, the numerical computations suggest that the two branches of solitary waves bifurcate from a finite-amplitude periodic waves at $c = c_{min}$, similar with the two-dimensional case [63, 71]. Using a linear elastic plate approximation, flexural-gravity solitary waves in infinite depth have also been computed in [58].

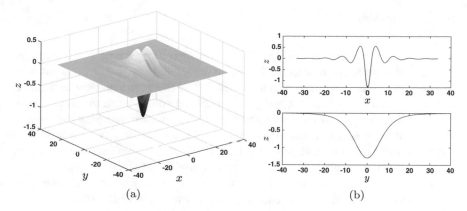

Fig. 4 Example of three-dimensional flexural-gravity depression solitary wave in infinite depth for $c = 1.276$ (**a**, left). Centrelines of the solution in x and y direction (**b**, right)

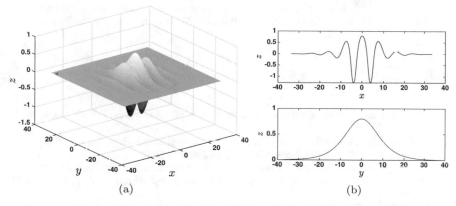

Fig. 5 Example of three-dimensional flexural-gravity elevation solitary wave in infinite depth for $c = 1.273$ (**a**, left). Centrelines of the solution in x and y direction (**b**, right)

In the gravity-capillary case, for strong surface tension $\sigma/\rho g h^2 > 1/3$ we found only fully-localised depression gravity-capillary solitary waves which are similar to the fully-localised solitary-wave solutions of the KP-I equation [60]. The waves no longer have decaying oscillations in the direction of propagation: the central depression is between two elevations which decay monotonically to zero as $x \to \infty$. It is also worth mentioning that in the flexural-gravity case a higher order KP equation was derived by Guyenne and Părău [36].

In all the cases it becomes numerically quite challenging to follow the branches for large-amplitude solutions to investigate the limiting configurations. We have investigated so far only symmetric waves in three dimensions, but based on the two-dimensional case and on a weakly nonlinear model equation for the gravity-capillary case [69], we expect to find non-symmetric solitary waves in three dimensions for the fully-nonlinear problems.

5 Conclusion

A number of numerical methods used for computing solitary waves in two and three dimensions have been reviewed. Different types of solitary waves obtained with these numerical methods have been discussed. While there are a number of similarities between the flexural-gravity and gravity-capillary waves, some differences have also been highlighted. In particular, the solitary waves with decaying oscillations bifurcate at $c = c_{\min}$ at a finite-amplitude in the flexural-gravity case in infinite depth, while they bifurcate at zero-amplitude in the gravity-capillary case. Another type of gravity-capillary solitary is found in finite depth when there is a strong surface tension, bifurcating at the long-wave limit $c = c_0$. Branches of symmetric and non-symmetric solitary and generalised solitary waves have also been presented.

Acknowledgements The authors are grateful to the Erwin Schrödinger International Institute for Mathematics and Physics, Vienna, for the support and hospitality during the 2017 *Nonlinear Water Waves—an Interdisciplinary Interface* workshop. This work was partially supported by EPSRC grants EP/J019305/1 for Emilian I. Pǎrǎu and EP/J019569/1 for Jean-Marc Vanden-Broeck.

References

1. M.J. Ablowitz, H. Segur, On the evolution of packets of water waves. J. Fluid Mech. **92**(4), 691–715 (1979)
2. B. Akers, P.A. Milewski, A model equation for wavepacket solitary waves arising from capillary-gravity flows. Stud. Appl. Math. **122**(3), 249–274 (2009)
3. T.R. Akylas, Envelope solitons with stationary crests. Phys. Fluids A **5**(4), 789–791 (1993)
4. T.R. Akylas, Three-dimensional long water-wave phenomena. Ann. Rev. Fluid Mech. **26**(1), 191–210 (1994)
5. T.R. Akylas, Y. Cho, On the stability of lumps and wave collapse in water waves. Philos. Trans. R. Soc. A **366**(1876), 2761–2774 (2008)
6. M.R. Alam, Dromions of flexural-gravity waves. J. Fluid Mech. **719**, 1–13 (2013)
7. D.J. Benney, G.J. Roskes, Wave instabilities. Stud. Appl. Math. **48**, 377–385 (1969)
8. K.M. Berger, P.A. Milewski, The generation and evolution of lump solitary waves in surface-tension-dominated flows. SIAM J. Appl. Math. **61**(3), 731–750 (2000)
9. B. Buffoni, M.D. Groves, S.-M. Sun, E. Wahlén, Existence and conditional energetic stability of three-dimensional fully localised solitary gravity-capillary water waves. J. Differ. Equ. **254**, 1006–1096 (2013)
10. B. Buffoni, M.D. Groves, E. Wahlén, A variational reduction and the existence of a fully localised solitary wave for the three-dimensional water-wave problem with weak surface tension. Arch. Rat. Mech. Anal. **228**, 773–820 (2018)
11. A.R. Champneys, J.-M. Vanden-Broeck, G.J. Lord, Do true elevation gravity-capillary solitary waves exist? A numerical investigation. J. Fluid Mech. **454**, 403–417 (2002)
12. D. Clamond, D. Dutykh, A. Durán, A plethora of generalised solitary gravity-capillary water waves. J. Fluid Mech. **784**, 664–680 (2015)
13. W. Craig, C. Sulem, Numerical simulation of gravity waves. J. Comput. Phys. **108**, 73–83 (1993)
14. A. Davey, K. Stewartson, On three-dimensional packets of surface waves. Proc. R. Soc. A **338**, 101–110 (1974)

15. F. Dias, G. Iooss, Capillary-gravity solitary waves with damped oscillations. Phys. D **65**(4), 399–423 (1993)
16. F. Dias, G. Iooss, Water-waves as a spatial dynamical system, in *Handbook of Mathematical Fluid Dynamics*, vol. 2 (North-Holland, Amsterdam, 2003), pp. 443–499
17. F. Dias, C. Kharif, Nonlinear gravity and capillary gravity waves. Annu. Rev. Fluid Mech. **31**, 301–346 (1999)
18. F. Dias, P. Milewski, On the fully-nonlinear shallow-water generalized Serre equations. Phys. Lett. A **374**(8), 1049–1053 (2010)
19. F. Dias, D. Menasce, J.-M. Vanden-Broeck, Numerical study of capillary-gravity solitary waves. Eur. J. Mech. B/Fluids **15**, 17–36 (1996)
20. V.D. Djordjevic, L.G. Redekopp, On two-dimensional packets of capillary-gravity waves. J. Fluid Mech. **79**, 703–714 (1977)
21. A.I. Dyachenko, E.A. Kuznetsov, M.D. Spector, V.E. Zakharov, Analytical description of the free surface dynamics of an ideal fluid (canonical formalism and conformal mapping). Phys. Lett. A **221**, 73–79 (1996)
22. L.K. Forbes, Surface waves of large amplitude beneath an elastic sheet. Part 1. High-order series solution. J. Fluid Mech. **169**, 409–428 (1986)
23. L.K. Forbes, Surface waves of large amplitude beneath an elastic sheet. Part 2. Galerkin solution. J. Fluid Mech. **188**, 491–508 (1988)
24. L.K. Forbes, An algorithm for 3-dimensional free surface problems in hydrodynamics. J. Comput. Phys. **82**, 330–347 (1989)
25. T. Gao, J.-M. Vanden-Broeck, Numerical studies of two-dimensional hydroelastic periodic and generalised solitary waves. Phys. Fluids **26**, 087101 (2014)
26. T. Gao, Z. Wang, J.-M. Vanden-Broeck. New hydroelastic solitary waves in deep water and their dynamics. J. Fluid Mech. **788**, 469–491 (2016)
27. T. Gao, Z. Wang, J.-M. Vanden-Broeck, On asymmetric generalized solitary gravity-capillary waves in finite depth. Proc. R. Soc. A **472**, 20160454 (2016)
28. T. Gao, J.-M. Vanden-Broeck, Z. Wang, Numerical computations of two-dimensional flexural-gravity solitary waves on water of arbitrary depth. IMA J. Appl. Math. **83**, 436–450 (2018)
29. A.G. Greenhill, Wave motion in hydrodynamics. Am. J. Math. **9**, 62–96 (1886)
30. R. Grimshaw, B. Malomed, E. Benilov, Solitary waves with damped oscillatory tails: an analysis of the fifth-order Korteweg-de Vries equation. Phys. D **77**, 473–485
31. M.D. Groves, S.-M. Sun, Fully localised solitary-wave solutions of the three-dimensional gravity-capillary water-wave problem. Arch. Rat. Mech. Anal. **188**, 1–91 (2008)
32. M.D. Groves, B. Hewer, E. Wahlén, Variational existence theory for hydroelastic solitary waves. C. R. Math. Acad. Sci. Paris **354**, 1078–1086 (2016)
33. P. Guyenne, E.I. Părău, Computations of fully nonlinear hydroelastic solitary waves on deep water. J. Fluid Mech. **713**, 307–329 (2012)
34. P. Guyenne, E.I. Părău, Finite-depth effects on solitary waves in a floating ice sheet. J. Fluids Struct. **49**, 242–262 (2014)
35. P. Guyenne, E.I. Părău, Forced and unforced flexural-gravity solitary waves. Proc. IUTAM **11**, 44–57 (2014)
36. P. Guyenne, E.I. Părău, Asymptotic modeling and numerical simulation of solitary waves in a floating ice sheet, in *Proceedings of 25th International Ocean Polar Engineering Conference (ISOPE 2015), Kona, Hawaii, 21–26 June 2015*, pp. 467–475
37. M. Hărăguş-Courcelle, A. Il'ichev, Three-dimensional solitary waves in the presence of additional surface effects. Eur. J. Mech. B/Fluids **17**(5), 739–768 (1998)
38. J.K. Hunter, J.-M. Vanden-Broeck, Solitary and periodic gravity-capillary waves of finite amplitude. J. Fluid Mech. **134**, 205–219 (1983)
39. A.T. Il'ichev, V.J. Tomashpolskii, Characteristic parameters of nonlinear surface envelope waves beneath an ice cover under pre-stress. Wave Motion **86**, 11–20 (2019)
40. G. Iooss, K. Kirchgässner, Bifurcation d'ondes solitaires en présence d'une faible tension superficielle. C. R. Acad. Sci. Paris Ser. 1 **311**, 265–268 (1990)

41. G. Iooss, K. Kirchgässner, Water waves for small surface tension: an approach via normal form. Proc. R. Soc. Edin. A **122**, 267–299 (1992)
42. G. Iooss, P. Kirrmann, Capillary gravity waves on the free surface of an inviscid fluid of infinite depth. Existence of solitary waves. Arch. Rat. Mech. Anal. **136**, 1–19 (1996)
43. B.B. Kadomtsev, V.I. Petviashvili, On the stability of solitary waves in weakly dispersing media. Sov. Phys. Dokl. **15**(6), 539–541 (1970)
44. B. Kim, T.R. Akylas, On gravity-capillary lumps. J. Fluid Mech. **540**, 337–351 (2005)
45. K. Kirchgässner, Nonlinear resonant surface waves and homoclinic bifurcation. Adv. Appl. Math. **26**, 135–181 (1988)
46. A. Korobkin, E.I. Părău, J.-M. Vanden-Broeck, The mathematical challenges and modelling of the hydroelasticity. Philos. Trans. Royal Soc. A. **369**, 2803–2812 (2011)
47. D.J. Korteweg, G. de Vries, On the change of form of long waves advancing in a rectangular canal and on a new type of long stationary waves. Philos. Mag. **36**, 422–433 (1895)
48. M. Longuet-Higgins, Capillary-gravity waves of solitary type on deep water. J. Fluid Mech. **200**, 451–470 (1989)
49. M.S. Longuet-Higgins, Capillary-gravity waves of solitary type and envelope solitons on deep water. J. Fluid Mech. **252**, 703–711 (1993)
50. P.A. Milewski, Three-dimensional localized solitary gravity-capillary waves. Commun. Math. Sci. **3**(1), 89–99 (2005)
51. P.A. Milewski, J.-M. Vanden-Broeck, Z. Wang, Dynamics of steep two-dimensional gravity-capillary solitary waves. J. Fluid Mech. **664**, 466–477 (2010)
52. P.A. Milewski, J.-M. Vanden-Broeck, Z. Wang, Hydroelastic solitary waves in deep water. J. Fluid Mech. **679**, 628–640 (2011)
53. P.A. Milewski, J.-M. Vanden-Broeck, Z. Wang, Steady dark solitary flexural gravity waves. Proc. R. Soc. A **469**, 20120485 (2013)
54. P.A. Milewski, Z. Wang, Three dimensional flexural-gravity waves. Stud. Appl. Math. **131**(2), 135–148 (2013)
55. E. Părău, F. Dias, Nonlinear effects in the response of a floating ice plate to a moving load. J. Fluid Mech. **460**, 281–305 (2002)
56. E.I. Părău, J.-M. Vanden-Broeck, Nonlinear two- and three- dimensional free surface flows due to moving disturbances. Eur. J. Mech. B/Fluids **21**, 643–656 (2002)
57. E.I. Părău, J.-M. Vanden-Broeck, Three-dimensional waves beneath an ice sheet due to a steadily moving pressure. Philos. Trans. R. Soc. A **369**, 2973–2988 (2011)
58. E.I. Părău, J.-M. Vanden-Broeck, Three-dimensional nonlinear waves under an ice sheet and related flows, in *Proceedings of 21st International Offshore Polar Engineering Conference (ISOPE-2011), Maui, 19–24 June 2011* (International Society of Offshore and Polar Engineers (ISOPE), Mountain View, 2011)
59. E.I. Părău, J.-M. Vanden-Broeck, M.J. Cooker, Nonlinear three-dimensional gravity-capillary solitary waves. J. Fluid Mech. **536**, 99–105 (2005)
60. E.I. Părău, J.-M. Vanden-Broeck, M.J. Cooker, Three-dimensional gravity-capillary solitary waves in water of finite depth and related problems. Phys. Fluids. **7**, 122101 (2005)
61. P.I. Plotnikov, J.F. Toland, Modelling nonlinear hydroelastic waves. Philos. Trans. R. Soc. A **369**, 2942–2956 (2011)
62. F. Smith, A. Korobkin, E. Parau, D. Feltham, V. Squire, Modelling of sea-ice phenomena. Philos. Trans. R. Soc. A **376**, 20180157 (2018)
63. O. Trichtchenko, E.I. Parau, J.-M. Vanden-Broeck, P. Milewski, Solitary flexural-gravity waves in three dimensions. Philos. Trans. R. Soc. A **376**(2129), 20170345 (2018)
64. J.-M. Vanden-Broeck, Elevation solitary waves with surface tension. Phys. Fluids A **3**, 2659–2663 (1991)
65. J.-M. Vanden-Broeck, *Gravity-Capillary Free-Surface Flows* (Cambridge University Press, Cambridge, 2010)
66. J.-M. Vanden-Broeck, F. Dias, Gravity-capillary solitary waves in water of infinite depth and related free-surface flows. J. Fluid Mech. **240**, 549–555 (1992)

67. J.-M. Vanden-Broeck, E.I. Părău, Two-dimensional generalised solitary waves and periodic waves under an ice sheet. Philos. Trans. R. Soc. A. **369**, 2957–2972 (2011)
68. Z. Wang, P.A. Milewski, Dynamics of gravity-capillary solitary waves in deep water. J. Fluid Mech. **708**, 480–501 (2012)
69. Z. Wang, J.-M. Vanden-Broeck, Multilump symmetric and nonsymmetric gravity-capillary solitary waves in deep water. SIAM J. Appl. Math. **75**, 978–998 (2015)
70. Z. Wang, J.-M. Vanden-Broeck, P.A. Milewski, Two-dimensional flexural-gravity waves of finite amplitude in deep water. IMA J. Appl. Math. **78**, 750–761 (2013)
71. Z. Wang, P.A. Milewski, J.-M. Vanden-Broeck, Computation of three-dimensional flexural-gravity solitary waves in arbitrary depth. Proc. IUTAM **11**, 119–129 (2014)
72. Z. Wang, J.-M. Vanden-Broeck, P.A. Milewski, Asymmetric gravity-capillary solitary waves on deep water. J. Fluid Mech. **759**, R2 (2014)
73. X. Xia, H.T. Shen, Nonlinear interaction of ice cover with shallow water waves in channels. J. Fluid Mech. **467**, 259–268 (2002)
74. T.S. Yang, T.R. Akylas, On asymmetric gravity-capillary solitary waves. J. Fluid Mech. **330**, 215–232 (1997)

A Method for Identifying Stability Regimes Using Roots of a Reduced-Order Polynomial

Olga Trichtchenko

Abstract For dispersive Hamiltonian partial differential equations of order $2N + 1$, $N \in \mathbb{Z}$, there are two criteria to analyse to examine the stability of small-amplitude, periodic travelling wave solutions to high-frequency perturbations. The first necessary condition for instability is given via the dispersion relation. The second criterion for instability is the signature of the eigenvalues of the spectral stability problem given by the sign of the Hamiltonian. In this work, we show how to combine these two conditions for instability into a polynomial of degree N. If the polynomial contains no real roots, then the travelling wave solutions are stable. We present the method for deriving the polynomial and analyse its roots using Sturm's theory via an example.

Keywords Spectral stability · High-frequency instabilities · Hamiltonian PDE · Dispersion · Sturm's theory

Mathematics Subject Classification (2000) Primary 35B35; Secondary 37K45

1 Introduction

Partial differential equations (PDEs) are used in a wide variety of applications to describe physical phenomena where this physical relevance imposes the requirement that the solutions to the PDEs are real. Moreover, if the description is of a closed system, there is usually an associated conservation of energy and the equations used are Hamiltonian. As more methodology for solving PDEs is developed [7], the natural question to ask is then how realistic are these solutions are and how likely we are to observe them in nature. Thus, analysing their stability also becomes important [1, 4, 5]. The purpose of this work is to present a simplified

O. Trichtchenko (✉)
Department of Physics and Astronomy, The University of Western Ontario, London, ON, Canada
e-mail: otrichtc@uwo.ca

© Springer Nature Switzerland AG 2019
D. Henry et al. (eds.), *Nonlinear Water Waves*, Tutorials, Schools, and Workshops in the Mathematical Sciences, https://doi.org/10.1007/978-3-030-33536-6_12

method for stability analysis, illustrated by an explicit example. We focus on high-frequency instabilities arising from spectral analysis of a perturbation of periodic travelling waves [3] and restrict our focus to stability of solutions of dispersive Hamiltonian equations. We show how working with the dispersion relation, we can methodically construct a parameter regime where there is only spectral stability with respect to particular perturbations and in the regions where we expect instability, we show what types of instabilities can arise.

In recent work [3], a method for establishing the presence of high-frequency instabilities of travelling wave solutions for both scalar PDEs as well as for systems of equations was described. In this method, there are two important conditions to consider:

1. collisions of eigenvalues of the spectral stability problem and
2. the signature of these eigenvalues.

Furthermore, it was shown that in order for the solutions to become unstable, the system had to admit waves travelling in different directions (bi-directional waves). In the follow-up work [11], the authors showed that a different way to meet the instability criteria, was for equations to contain what is referred to as a generalised resonance. An equation contains a resonance if there is a certain set of parameters for which travelling wave solutions are predominantly composed of at least two distinct frequencies which can travel at the same speed. Physically, this implies that there are at least two different forces that can influence the travelling waves that are of the same order of magnitude. For example, if we are considering water waves, then these waves are in a **resonant regime** if surface tension and gravity are competing forces of the same order of magnitude. The result is that the travelling wave profiles contain two different prominent modes, otherwise referred to as Wilton ripples [10, 12].

If we restrict ourselves to scalar, dispersive and Hamiltonian PDEs where the solution u depends on one spatial and one time variable, i.e. $u = u(x, t)$ with a period L and up to $2N + 1$ derivatives, then it has been shown [3] that all we need is a polynomial dispersion relation $\omega(k)$ of order $2N + 1$ to describe both of the necessary conditions for instability. In [6], it was shown that the two necessary conditions for instability can be collapsed into one criterion on the roots of a polynomial of order N to be in an interval I defined in Sect. 2. This greatly simplifies the analysis, leading to closed-form results for stability regions of specific PDEs.

This work presents a method for the single criteria for instability of periodic travelling wave solutions to a dispersive, Hamiltonian PDE using an example with three competing terms. The formulation and underlying theory is described in Sect. 2. Working with the dispersion relation, we show the general methodology for the stability analysis in Sect. 3. Section 4 explicitly shows how to implement the method via an example, demonstrating how to construct the coefficients systematically and use Sturm's theory to analyse the roots of the reduced polynomial. In Sect. 5, figures of the stability and instability regions are shown and we conclude in Sect. 6.

2 Summary of Stability Theory

Consider a scalar Hamiltonian PDE of the form

$$u_t = \partial_x \frac{\delta H}{\delta u}, \tag{2.1}$$

where the function $u = u(x, t)$ describes a periodic travelling wave, with H the Hamiltonian and $\frac{\delta H}{\delta u}$ a variational derivative. More specifically $u(x, t)$ is a solution of

$$u_t = \sum_{n=1}^{N} C_{2n+1} \frac{\partial^{2n+1} u}{\partial x^{2n+1}} + f(u, u_x, \ldots, u_{(2N)x})_x, \tag{2.2}$$

where N is positive integer and $\frac{\partial^{2n+1} u}{\partial x^{2n+1}}$ are $2n + 1$ (odd) derivatives up to order $2N + 1$ with the nonlinearity f that can depend on u as well as its derivatives up to order $2N$ (denoted as $u_{(2N)x}$), keeping the overall system dispersive. For ease, we consider the equation with real coefficients C_{2n+1}. We obtain the dispersion relation $\omega(k)$ if we let $u(x, t) \sim e^{ikx - i\omega t}$ with k a Fourier mode, and substitute into (2.2) to obtain

$$\omega(k) = \sum_{n=1}^{N} (-1)^{(n+1)} C_{2n+1} k^{2n+1}. \tag{2.3}$$

Furthermore, if we restrict the space of solutions $u(x, t)$ to periodic, travelling waves moving at speed V such that $u(x, t) \to u^{(0)}(x - Vt)$, then we can write (2.2) in the travelling frame of reference and consider the steady-state equation

$$V u_x + \sum_{n=1}^{N} C_{2n+1} \frac{\partial^{2n+1} u}{\partial x^{2n+1}} + f(u, u_x, \ldots, u_{(2N)x})_x = 0, \tag{2.4}$$

and setting $x \to x - Vt$ from now on. Despite restricting the space of solutions to travelling waves $u^{(0)}(x)$, we can still gather information about the time dependence by perturbing about this steady-state with a small perturbation governed by δ, i.e.

$$u(x, t) = u^{(0)}(x) + \delta \bar{u}^{(1)}(x, t)$$

$$= u^{(0)}(x) + \delta e^{\lambda t} u^{(1)}(x). \tag{2.5}$$

We have made an assumption about the time dependence of the perturbation by introducing $\lambda \in \mathbb{C}$. Recall that $u^{(0)}(x)$ is periodic of period L (for convenience, $L = 2\pi$) [3]. We allow the perturbations to be of any period, but bounded in space using the Fourier-Floquet expansion

$$u^{(1)}(x) = e^{i\mu x} \sum_{m=-M}^{M} b_m e^{imx}, \tag{2.6}$$

with $\mu \in \mathbb{R}$ the Floquet parameter governing the period of the perturbation and a Fourier mode $m \in \mathbb{Z}$ [2]. We note that this perturbation can grow exponentially in time if $\mathrm{Re}(\lambda) > 0$, where $\lambda = \lambda(\mu + m)$ depends on the Fourier-Floquet modes m and μ. For solutions with $|u^{(0)}(x)| = O(\epsilon)$ with $\epsilon \to 0$,

$$\lambda(\mu + m) = i(m + \mu)V - i\omega(m + \mu), \tag{2.7}$$

if we consider $O(\delta)$ term when substituting (2.5) and (2.6) into (2.2), staying in the travelling frame of reference.

For ease of notation, we introduce the dispersion relation Ω in the travelling frame of reference as $\Omega(m + \mu) = \omega(m + \mu) - (m + \mu)V$ with $\lambda(\mu + m) = -i\Omega(m + \mu)$. Since λ is purely imaginary when we consider the linear regime, the perturbation will not grow exponentially in time and thus $u^{(0)}(x)$ is spectrally stable. However, as the nonlinearity is increased with increasing ϵ, the eigenvalues which depend continuously on the amplitude of the solution will change and may develop some non-zero real part. Since the equation is Hamiltonian, they will do so symmetrically in the complex plane to conserve the energy, keeping the solution real. The possible configurations of the symmetries in eigenvalues are shown in Fig. 1. In order to leave the imaginary axis and develop instability, the eigenvalues first have to collide in order to maintain the symmetry of the equation. In Fig. 1, even if eigenvalues move and collide, they do not necessarily leave the imaginary axis as shown in the left panel. This implies a necessary condition for instability is

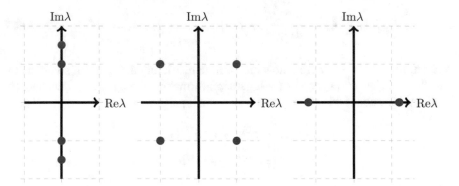

Fig. 1 Three different configurations of the smallest number of eigenvalues λ of the spectral stability problem of a Hamiltonian system, showing the symmetry about the real and imaginary axes. On the left (in blue), is the stable regime. The centre and right panel are the unstable regimes (in red)

collisions of eigenvalues for different modes m and n in a perturbation given by

$$\lambda(\mu + m) = \lambda(\mu + n). \tag{2.8}$$

Also in the linear regime (considering the $O(\delta)$ term when substituting (2.5) into (2.2) with $|u^{(0)}(x)| \to 0$), we can explicitly write the Hamiltonian of the system as

$$H_{\text{lin}} = \int_0^L \frac{1}{2} \left(\sum_{n=1}^N (-1)^n C_{2n+1} (u_{nx}^{(1)})^2 + V (u^{(1)})^2 \right) dx, \tag{2.9}$$

with

$$0 = \partial_x \frac{\delta H_{\text{lin}}}{\delta u^{(1)}}. \tag{2.10}$$

An unstable solution has to conserve energy given by (2.9). This implies that for a collision of eigenvalues arising from two different modes, for every mode that is contributing positively to the Hamiltonian, there needs to be a negatively contributing mode as well. This contribution of eigenvalues to the Hamiltonian (known as their **signature**) is simply given by the sign of the Hamiltonian. The signature is derived from (2.9) by substituting $u^{(1)} \sim e^{i(\mu+m)x}$ to obtain

$$\text{sign}(H_{\text{lin}}) = \text{sign} \left(\sum_{m=1}^N (-1)^m C_{2m+1} (i(\mu + m))^{2m} + V \right). \tag{2.11}$$

Using the definition of the dispersion relation in the moving frame and dividing by i, we can write the sign of the Hamiltonian as

$$\text{sign}(H_{\text{lin}}) = \text{sign} \left(\frac{\Omega(\mu + m)}{\mu + m} \right), \tag{2.12}$$

where we have used (2.3) and the definition of the dispersion relation incorporating the travelling frame of reference. With more algebra described in [3, 6], we can introduce s which will govern if two colliding eigenvalues for modes m and n will have opposing signature as

$$s = (\mu + m)(\mu + n) < 0. \tag{2.13}$$

To reduce the number of unknowns in (2.13), we set $(\mu + m) \to \mu$ therefore letting $n \to (n - m)$, shifting the focus instead on the difference in Fourier modes of the perturbation. This implies that if we wish to consider when the periodic travelling wave solutions are unstable to perturbations of the form shown in (2.6), then we need examine the **collision condition**

$$\lambda(\mu) = \lambda(\mu + n) \tag{2.14}$$

as well as the corresponding combination of **signatures of colliding eigenvalues** given by

$$s = \mu(\mu + n) \tag{2.15}$$

In the following sections, we show this can be further simplified to one condition using a reduced order polynomial of degree N and examine where the polynomial has real roots thereby meeting the necessary conditions for instability.

3   General Methodology

In general, if we are given a polynomial with $p(\mu) = \mu^N$ with N odd (for example one term in a dispersion relation), then a collision of eigenvalues is of the form

$$p(\mu + n) - p(\mu) = 0. \tag{3.1}$$

Setting $s = \mu(\mu+n)$, we can equivalently write the collision condition as a reduced-order polynomial $q(s, n)$ of order $\frac{N-1}{2}$ that is indirectly dependent on the Floquet parameter μ as

$$q(s, n) = \sum_{i=0}^{\frac{N-1}{2}} a_{i,N-2i} s^i n^{N-2i}. \tag{3.2}$$

The coefficients can be computed recursively as

$$a_{i,j} = \begin{cases} \binom{N}{j} & \text{for } i = 0, \, j = 2, \ldots, N, \\ a_{i-1,j+1} - a_{i,j+1} & \text{for } i = 1, \ldots, \frac{N-1}{2}, \, j = 1, \ldots, N - 2i, \\ 0 & \text{otherwise.} \end{cases} \tag{3.3}$$

Rewriting the collision condition as a signature condition is always possible as shown by Kollar et al. in [6]. In the following section we will focus on the simplicity of constructing this polynomial for the signature. The main consequence of being able to rewrite the polynomial of lower order, is that it simplifies the equation and the number of roots we have to consider. From (2.15), we can solve for the Floquet parameter as

$$\mu = \frac{1}{2} \left(-n \pm \sqrt{n^2 + 4s} \right). \tag{3.4}$$

To satisfy both the collision condition and signature condition for instability while maintaining that perturbations are bounded in space, we need the roots of (3.4) to be real and for the signatures to remain opposite, i.e.

$$-\frac{n^2}{4} < s < 0. \tag{3.5}$$

Checking that the roots of a polynomial are within a certain interval I, in this case given by (3.5), becomes a relatively straightforward procedure and is in some respect easier than computing exact roots. This can be done using Sturm's theory [8, 9] via a sequence of polynomials (sometimes known as a Sturm chain). Given a polynomial $g(x) = g_0(x)$ of degree N with real coefficients, a sequence of polynomials of decreasing order is constructed by using the following criteria

$$g_1(x) = \frac{\partial}{\partial x} g_0(x) \text{ and} \tag{3.6}$$

$$g_n(x) = -\left(g_{n-2}(x) - g_{n-1}(x)\frac{g_{n-2}(x)}{g_{n-1}(x)}\right) = -\text{Rem}(g_{n-2}(x), g_{n-1}(x)) \tag{3.7}$$

where $\frac{g_{n-2}(x)}{g_{n-1}(x)}$ is a polynomial quotient and $\text{Rem}(g_{n-2}(x), g_{n-1}(x))$ is the remainder. The sequence terminates at $n = N$ when the last term is a constant and therefore independent of x. If we are interested in how many real roots r_n occur in the interval $I = (a_i, a_f)$, where a_i and a_f are not themselves roots, then we need to examine the difference in the number of sign changes of the polynomials evaluated at the endpoints of the interval (as shown in (3.5), in this case $a_i = -n^2/4$ and $a_f = 0$). To obtain the number of real roots in the interval, we subtract the number of sign changes at a_f from the number of sign changes at a_i.

To summarise, in order to analyse spectral stability of periodic travelling waves of (2.4) to high-frequency instabilities of the form given by (2.6), we must

1. Write the dispersion relation ω given by the general form in (2.3).
2. Compute the travelling wave speed V for a non-trivial solution.
3. Solve for the polynomial that governs the collision condition of the form (2.14).
4. Reduce the order of the polynomial by substituting $s = \mu(\mu + n)$.
5. Generate the Sturm sequence of polynomials using (3.7).
6. Compute the number of roots in I by examining the number of sign changes in the Sturm sequence of polynomials at each end point and noting the difference.

If the result is that we have no real roots contained in I, then the periodic travelling waves are spectrally stable to high-frequency perturbations. In order to show how this method works, we proceed with an example.

4 Example

In this section we examine an equation of the form

$$u_t + \alpha u_{3x} + \beta u_{5x} + \gamma u_{7x} + f(u)_x = 0, \tag{4.1}$$

where α, β and γ are real coefficients and the subscripts represent the number of derivatives of $u(x, t)$ and go through the process outlined in Sect. 3 to compute the regions of stability, referring to step number in parentheses. In this section, we will keep these as variables however in practice, they are defined by the scaling in the partial differential equation that is being considered. We begin by introducing a travelling frame of reference, moving with speed V and considering a steady-state solution

$$\alpha u_{3x} + \beta u_{5x} + \gamma u_{7x} + f(u)_x + V u_x = 0. \tag{4.2}$$

The dispersion relation (step 1 in the process) of this equation is given by

$$\omega = -\alpha k^3 + \beta k^5 - \gamma k^7. \tag{4.3}$$

Linearizing about a small amplitude solution with $u^{(0)} = \epsilon e^{ikx}$ (where $f(u_x^{(0)}) \approx 0$), we obtain

$$\alpha(ik)^3 + \beta(ik)^5 + \gamma(ik)^7 + V(ik) = 0, \tag{4.4}$$

or

$$-\alpha k^2 + \beta k^4 - \gamma k^6 + V = 0. \tag{4.5}$$

If we assume the solution we are linearising about is 2π periodic, we can show it is symmetric and without loss of generality we can set $k = 1$. This gives $V_0 = \alpha - \beta + \gamma$ (completing step 2) as a bifurcation point from which we can compute non-trivial solutions $u^{(0)}(x)$ travelling at speed V_0. We will sub in for $V = V_0$ in the equations from now on.

The polynomial in terms of (μ, n) (step 3) for the collision condition is given by

$$p(\mu, n) = \gamma(\mu + n)^7 - \beta(\mu + n)^5 + \alpha(\mu + n)^3 - \gamma\mu^9 + \beta\mu^5 - \alpha\mu^3 - (\alpha - \beta + \gamma)n. \tag{4.6}$$

The above can be simplified if we set $s = \mu(\mu + n)$. In order to do this, we first note that we can use binomial theorem gives us the polynomial expansion

$$(\mu + n)^N = \sum_{k=0}^{N} \binom{N}{k} \mu^{N-k} n^k. \tag{4.7}$$

Table 1 Coefficients from the binomial theorem in a Pascal's triangle

$$
\begin{array}{ccccccccccccccc}
& & & & & & & 1 & & & & & & & \\
& & & & & & 1 & & 1 & & & & & & \\
& & & & & 1 & & 2 & & 1 & & & & & \\
& & & & 1 & & 3 & & 3 & & 1 & & & & \\
& & & 1 & & 4 & & 6 & & 4 & & 1 & & & \\
& & 1 & & 5 & & 10 & & 10 & & 5 & & 1 & & \\
& 1 & & 6 & & 15 & & 20 & & 15 & & 6 & & 1 & \\
1 & & 7 & & 21 & & 35 & & 35 & & 21 & & 7 & & 1
\end{array}
$$

The coefficients from the binomial theorem can be computed via Pascal's triangle where each row represents coefficients in a polynomial of degree $N = 0, \cdots, 7$ shown in Table 1 and obtain the collision condition as

$$
\begin{aligned}
p(\mu, n) =\,& \gamma (7\mu^6 n + 21\mu^5 n^2 + 35\mu^4 n^3 + 35\mu^3 n^4 + 21\mu^2 n^5 + 7\mu n^6 + n^7) \\
& - \beta \left(5\mu^4 n + 10\mu^3 n^2 + 10\mu^2 n^3 + 5\mu n^4 + n^5\right) + \alpha(3\mu^2 n + 3\mu n^2 + n^3) \\
& - (\alpha - \beta + \gamma()n = 0.
\end{aligned}
\tag{4.8}
$$

Just as Pascal's triangle provides an easy way to compute the coefficients of $(\mu + n)^N$ in (3.1), we can use a triangular construction to find the coefficients of $q(s, n)$ in (3.2). To begin, create a table whose N columns are the coefficients of $(\mu+n)^N - \mu^N$ beginning with the coefficient of n^N and ending with the coefficient of n^1. Row 2 begins with a zero one place to the left of the first column in row 1. Subsequent elements in row 2 are found by computing the difference between row 1 and row 2 in the previous column. This procedure is repeated until the final row which will have just two elements. The coefficients in the reduced polynomial for the signature (that is the polynomial which depends on $s = \mu(\mu+n)$) are the first non-zero values in each row (circled in Tables 2, 3 and 4). They are given in increasing order of s as labelled in the right-most row. That is, row 1 gives the coefficient of $s^0 n^N$ and row $(N + 1)/2$ gives the coefficient of $s^{(N-1)/2} n^1$. Tables 2, 3 and 4 show this process explicitly for $N = 7, 5, 3$ respectively.

Finally, combining the results from the Tables 2, 3 and 4, the polynomial for the signature condition (step 4 in the process) is

$$
\begin{aligned}
q(s, n) =\,& -\gamma(n^6 + 7n^4 s + 14n^2 s^2 + 7s^3) + \beta(n^4 + 5n^2 s + 5s^2) \\
& - \alpha(n^2 + 3s) + \alpha - \beta + \gamma
\end{aligned}
\tag{4.9}
$$

Table 2 Tabular computation of $(\mu + n)^7 - \mu^7 = n^7 + 7sn^5 + 14s^2n^3 + 7s^3n$

n^7	n^6	n^5	n^4	n^3	n^2	n^1	
①	7	21	35	35	21	7	s^0
↓	↓	↓	↓	↓	↓		
0 →	⑦ →	14 →	21 →	14 →	7		s^1
	↓	↓	↓	↓			
	0 →	⑭ →	7 →	7			s^2
		↓					
		0 →	⑦				s^3

The coefficients of the reduced polynomial in terms of $s = \mu(\mu + n)$ are given by the circled terms. Downward arrows (in blue) indicate subtraction and arrows to the right (in black) indicate the result of the subtraction

Table 3 Tabular computation of $(\mu + n)^5 - \mu^5 = n^5 + 5sn^3 + 5s^2n$

n^5	n^4	n^3	n^2	n^1	
①	5	10	10	5	s^0
↓	↓	↓			
0 →	⑤ →	5 →	5		s^1
	↓				
	0 →	⑤			s^2

The coefficients of the reduced polynomial in terms of $s = \mu(\mu + n)$ are given by the circled terms. Downward arrows (in blue) indicate subtraction and arrows to the right (in black) indicate the result of the subtraction

Table 4 Tabular computation of $(\mu + n)^3 - \mu^3 = n^3 + 3sn$

n^3	n^2	n^1	
① →	3 →	3	s^0
↓			
0 →	③		s^1

The coefficients of the reduced polynomial in terms of $s = \mu(\mu + n)$ are given by the circled terms. Downward arrows (in blue) indicate subtraction and arrows to the right (in black) indicate the result of the subtraction

We can analyse the roots of (4.9) using Sturm's theory by constructing a sequence of polynomials (this is step 5) in s of the form in (3.7) as

$$p_1(s) = -\gamma(7n^4 + 28n^2 s + 21s^2) + \beta(5n^2 + 10s) - 3\alpha \tag{4.10}$$

$$p_2(s) = -\frac{2s}{63\gamma}\left(49\gamma^2 n^4 - 35\beta\gamma n^2 - 63\alpha\gamma + 25\beta^2\right)$$

$$-\frac{1}{63\gamma}\left(35\gamma^2 n^6 - 42\beta\gamma n^4 - 21\alpha\gamma n^2 + 25\beta^2 n^2 - 15\alpha\beta + 63\alpha\gamma - 63\beta\gamma + 63\gamma^2\right) \tag{4.11}$$

$$p_3(s) = -\frac{63\gamma}{4(49\gamma^2 n^4 - 35\beta\gamma n^2 - 63\alpha\gamma + 25\beta^2)^2}\left(49\gamma^4 n^{12} - 196\beta\gamma^3 n^{10} - 98\alpha\gamma^3 n^8\right.$$

$$+322\beta^2\gamma^2 n^8 - 126\alpha\beta\gamma^2 n^6 + 1274\alpha\gamma^3 n^6 - 200\beta^3\gamma n^6 - 1274\beta\gamma^3 n^6 + 1274\gamma^4 n^6$$

$$+441\alpha^2\gamma^2 n^4 - 210\alpha\beta^2\gamma n^4 - 1176\alpha\beta\gamma^2 n^4 + 125\beta^4 n^4 + 1176\beta^2\gamma^2 n^4$$

$$-1176\beta\gamma^3 n^4 + 630\alpha^2\beta\gamma n^2 - 2646\alpha^2\gamma^2 n^2 - 250\alpha\beta^3 n^2 + 1050\alpha\beta^2\gamma n^2$$

$$+2646\alpha\beta\gamma^2 n^2 - 2646\alpha\gamma^3 n^2 - 1050\beta^3\gamma n^2 + 1050\beta^2\gamma^2 n^2 - 756\alpha^3\gamma$$

$$+225\alpha^2\beta^2 + 1890\alpha^2\beta\gamma - 1323\alpha^2\gamma^2 - 500\alpha\beta^3 - 1890\alpha\beta^2\gamma + 4536\alpha\beta\gamma^2$$

$$\left.-2646\alpha\gamma^3 + 500\beta^4 - 500\beta^3\gamma - 1323\beta^2\gamma^2 + 2646\beta\gamma^3 - 1323\gamma^4\right) \tag{4.12}$$

Despite the length of the expressions in (4.9)–(4.12), their sign changes are easy to evaluate for particular α, β, γ and $s \in (-1/4, 0)$. For ease, Table 5 shows the sign changes for $\alpha = 1$, $\beta = 1/4$ and $\gamma = 0$ for $n = 1, 2, 3, 4$ which are in complete

Table 5 The stability results with $\alpha = 1$, $\beta = 1/4$ and $\gamma = 0$ and $n = 1, 2, 3, 4$ (note this is a singular case of (4.1))

$n = 1$	sign($p_j(-n^2/4)$)	sign($p_j(0)$)	$n = 2$	sign($p_j(-n^2/4)$)	sign($p_j(0)$)
$p_0(s)$	+	+	$p_0(s)$	−	+
$p_1(s)$	−	−	$p_1(s)$	−	+
$p_2(s)$	+	+	$p_2(s)$	+	+
Sign changes	2	2	Sign changes	1	0
$n = 3$	sign($p_j(-n^2/4)$)	sign($p_j(0)$)	$n = 4$	sign($p_j(-n^2/4)$)	sign($p_j(0)$)
$p_0(s)$	−	+	$p_0(s)$	+	+
$p_1(s)$	+	+	$p_1(s)$	+	+
$p_2(s)$	+	+	$p_2(s)$	+	+
Sign changes	2	0	Sign changes	0	0

By subtracting the number of sign changes at $s = -n^2/4$ from the ones at $s = 0$ (subtract the total in column 3 from column 2) for each n, we get the number of real roots in that interval. Instability is possible for $n = 2$ and $n = 3$ since there are roots for which $s \in I = (-n^2/4, 0)$. We can also conclude the equation is stable to perturbations with $n = 1$ and $n \geq 4$

agreement with results in [11] (this is the final step in the process, step 6). They imply that the perturbations with $n \geq 4$ are stable as is the perturbation for $n = 1$ since there are no real roots. Note that in cases where $p_j(s) = 0$, we must consider the limit as s approaches the value 0 or $-n^2/4$ from the correct side to match with the condition in (3.5).

5   Stability Results

Figures 2, 3 and 4 show in more detail the stable and unstable regions in two-dimensions for PDEs with only one free parameter (setting one of the parameters in the PDE to zero). In Fig. 2, $\alpha = 0$, $\beta = 1$ and γ is a free parameter. The region

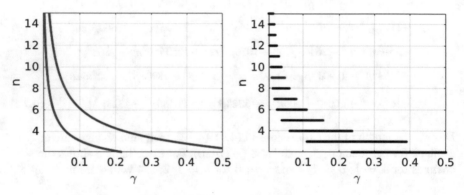

Fig. 2 Reduction to the two-dimensional system with $\alpha = 0$, giving the instability results for $u_t = V u_x + \beta u_{5x} + \gamma u_{7x} +$ nonlinearity

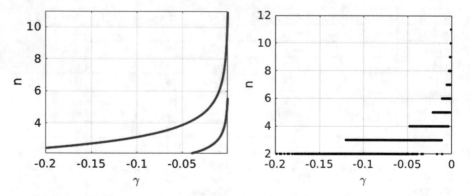

Fig. 3 Reduction to the two-dimensional system with $\beta = 0$, giving the instability results for $u_t = V u_x + \alpha u_{3x} + \gamma u_{7x} +$ nonlinearity

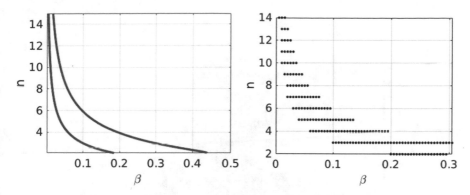

Fig. 4 Reduction to the two-dimensional system with the singular limit when $\gamma = 0$, giving the instability results for $u_t = V u_x + \alpha u_{3x} + \beta u_{5x} +$ nonlinearity

bounded below by the blue line and above by the red line is where we can have instability and outside of these curves is where the small amplitude solutions are stable with respect to the instabilities considered in this work. In the plot on the right in Fig. 2, the dots show the unstable regime for integer values of n where (4.9) has roots in the interval $(-n^2/4, 0)$. We see that as γ decreases, the instabilities occur for larger n, indicating the difference in Fourier modes of colliding eigenvalues. Figure 3 gives the stability regions for $\alpha = 1$, $\beta = 0$ and γ as a free parameter. In this case, only $\gamma < 0$ leads to instabilities, but the pattern is similar to the previous figure. Figure 4 gives the results previously computed in [11] where once again with decreasing β, the instabilities have an increasing n.

Figure 5 summarises the full stability results for the general PDE (2.2) with $\gamma = 1$, which is simply a rescaling of the full equation and does not reduce the degrees of freedom. The regions between the blue and red lines are possible regions of instability. For clarity, points in the lower plot of Fig. 5 show possible regions of instability and the white space gives the regimes for spectrally stable periodic travelling wave solutions to (2.2). This plot shows that most of the regimes of (2.2) are stable.

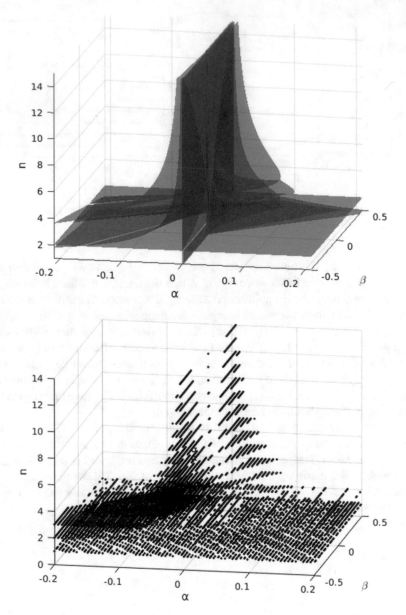

Fig. 5 On the top, the regions bounded by red and blue curves are those where instabilities can arise. On the bottom, the dots represent possible unstable regions for discrete values of n. For both figures, the equations were normalised such that $\gamma = 1$

6 Conclusion

In this work, we describe a systematic way to fully characterise spectral stability regions of travelling wave solutions of a dispersive, Hamiltonian PDE subject to high-frequency instabilities. This method shows explicitly how two necessary conditions can be merged into one and a systematic way to analyse the reality of its roots. It relies on reducing the polynomial derived from the dispersion relation describing collisions of eigenvalues of degree $2N + 1$, to a polynomial for the signature condition of degree N. This polynomial can be constructed using a triangle of coefficients as is illustrated using an example of a PDE containing three linear dispersive terms with general coefficients. If this reduced-order polynomial has roots in a given interval $I = (-n^2/4, 0)$, which can be determined using Sturm's theory, then the necessary criteria for an instability is met. This methodology can be used on any dispersive, Hamiltonian partial differential equation. Sturm's theory has also been implemented in Maple and can be accessed through the commands sturm and sturmseq.

There are two drawbacks to this method. One is that it can only be used if the sign of the Hamiltonian is definite, hence the restriction to high-frequency instabilities is made. It also relies on the underlying equations having a Hamiltonian and hence a fourfold symmetry in the complex eigenvalue plane. Since many physical systems are Hamiltonian, there is a large number of applications of this method (for more examples, see [3]), which also includes Euler equations describing water waves.

Acknowledgements We thank B. Deconinck, R. Kollár and D. Ambrose for insightful discussions. We wish to thank Casa Mathemática Oaxaca, the Erwin Schrödinger Institute and Institute for Computational and Experimental Research in Mathematics (ICERM) for their hospitality during the development of the ideas for this work.

References

1. J.C. Bronski, M.A. Johnson, T. Kapitula, An index theorem for the stability of periodic travelling waves of Korteweg–de Vries type. Proc. R. Soc. Edinb. Sect. A Math. **141**(6), 1141–1173 (2011)
2. B. Deconinck, J.N. Kutz, Computing spectra of linear operators using the Floquet-Fourier-Hill method. J. Comput. Phys. **219**, 296–321 (2006)
3. B. Deconinck, O. Trichtchenko, High-frequency instabilities of small-amplitude solutions of Hamiltonian PDEs, in *DCDS-A. 37* (2017), pp. 1323–1358
4. V.M. Hur, M.A. Johnson, Modulational instability in the Whitham equation for water waves. Stud. Appl. Math. **134**(1), 120–143 (2015)
5. R. Kollár, P.D. Miller, Graphical Krein signature theory and Evans–Krein functions. SIAM Rev. **56**(1), 73–123 (2014)
6. B. Kollár, B. Deconinck, O. Trichtchenko, Direct characterization of spectral stability of small amplitude periodic waves in scalar Hamiltonian problems via dispersion relation. SIAM J. Math. Anal. **51**(4), 3145–3169 (2019)
7. P.J. Olver, *Introduction to Partial Differential Equations* (Springer, Berlin, 2014)

8. C. Sturm, *Mémoire sur la résolution des équations numériques* (1835)
9. J.M. Thomas, Sturm's theorem for multiple roots. Natl. Math. Mag. **15**(8), 391–394 (1941)
10. O. Trichtchenko, B. Deconinck, J. Wilkening, The instability of Wilton ripples. Wave Motion **66**, 147–155 (2016)
11. O. Trichtchenko, B. Deconinck, R. Kollár, Stability of periodic travelling wave solutions to the Kawahara equation. SIAM J. Appl. Dyn. Syst. **17**(4), 2761–2783 (2018)
12. J.R. Wilton, On ripples. Philos. Mag. Ser. 6 **29**(173), 688–700 (1915)

Printed in the United States
By Bookmasters